WET CHEMISTRY ANALYSIS - BOOK 1 (REVISED)

William Lipps

Williamlipps.org

Forward

This book is intended for you. It is not intended to be overly technical or a "scientific" writing. It is intended to be an introduction to wet chemistry analysis. Wet chemistry is a fading art. As instrumentation improves, the analysis is simplified to the point that the analyst need not understand the chemistry behind the testing they are doing. Routine testing should be easy, and if an instrument or testing device provides an accurate result at the push of a button that's OK. However, matrices vary and at the present state of technology it is just not possible to have an accurate sensor for everything. While methods need to be standardized, they also need to be modified when they incorrectly measure an analyte or a concentration in a particular matrix. Before you can modify a method, you must have an idea how it works.

Table of Contents

Introduction to Wet Chemical Analysis 7

Necessary tools for use in any environmental laboratory 12

    Balance 12

    Volumetric Flasks 16

    Burettes 19

    Volumetric pipettes 21

    Air displacement pipettes 23

    Beakers and flasks 24

    PH / ISE meters 25

    Colorimeters 26

Quality Assurance / Quality Control 27

    Quality Control 28

    Calibration 36

    Accuracy and Precision 38

Chemical Analysis of things the way they are not 40

Environmental Sampling 43

    Total Metals 46

    Dissolved Metals 46

    Total Nitrogen (TKN), Total Phosphorus, Ammonia - Nitrogen, Nitrate + Nitrite - Nitrogen 47

    Phosphate 47

    Phenolics 48

    Cyanide 48

Method Development 49

Example of Applied Methodology - Oilfield water 52
Sampling oilfield waters 54

Laboratory Analysis...57

Volumetric Analysis..58
Titration..58
Neutralization Reactions 62
Standardization of titration reagents............................67
Ammonia Nitrogen by titration 69
Alkalinity Analysis by titration.......................................74
Other examples of titration 76

The Ion Selective Electrode..78
Routine operation of ISE direct potentiometry 85
Determination of Ammonia by ISE direct potentiometry 87
Fluoride Analysis by ISE 90
Nitrate by Ion Selective Electrode 93

Colorimetry 97
Explanation of Beer's Law..98
Photometric analysis of Ammonia 107
Ammonia Analysis by Manual Spectrophotometer................109
Nitrite Nitrogen by Manual Colorimetry 112
Nitrate plus Nitrite by manual cadmium reduction and manual colorimetry..112
Residual Chlorine by Manual Colorimetry 115
Colorimetric Test Kits and Pre-packaged reagents 117

Automated methods of analysis 118
What is automation? 120
Should you automate?...126

Continuous Flow Analyzers..129
History of laboratory automation of wet chemical analysis ..130
Flow Injection Analysis 133
Segmented Flow Analysis 139
Some similarities and differences between SFA and FIA.......150
Interpreting a Flow method diagram 153
Summary of SFA and FIA..156

The Discrete Analyzer..159

Comparison of Discrete and Continuous Flow Analyzers..164
Choosing between Continuous Flow or Discrete Analyzers 168
    Different Types of Discrete Analyzers 170
    Flow-cell designs 170
    In reaction cuvette designs...173

Ion Chromatography 175
    Interferences with ion chromatography 179
    Advantages of ion chromatography 180
    Summary of ion chromatography..187

Conclusion...189

Introduction to Wet Chemical Analysis

Wet chemistry represents various chemical techniques that are used to determine the concentration of an analyte, or pollutant, by reactions with aqueous reagents. Wet chemistry involves measuring, weighing, and mixing reagents with sample solutions to form a precipitate that can be weighed or an aqueous product that can be measured either by titration, or by instrumentation.

Gravimetric Analysis may be simply evaporation of an aqueous sample to measure the residue, filtration and measurement of the mass collected on the filter, or it may involve the addition of chemicals to form a precipitate that can be weighed. Gravimetric techniques can be very precise but are also time consuming. Many environmental tests, such as determination of dissolved and suspended solids are gravimetric techniques.

Volumetric Analysis determines an analyte dissolved in aqueous solution by the addition of a known quantity of another solution containing a known amount of reactant. The reaction product between the two solutions is visually or electrochemically measured at the equivalence point of the reaction. The analyst is able to calculate the concentration of the analyte. A volumetric analysis is also known as a *titration*. Titrations are faster than gravimetric analysis and preferred if there is a volumetric alternative to the gravimetric test. Examples of titrations in environmental chemistry are determination of alkalinity by addition of acid, or determination of chloride by addition of silver ion.

Colorimetric analysis determines the concentration of an analyte dissolved in aqueous solution by the addition of chemical reagents that cause a color reaction between the chemical and the analyte. The intensity of the color is proportional to concentration. The analyst prepares a series of solutions with increasing concentration of the analyte and adds the color forming reagent to these solutions, called standards, at the same time as adding the color reagent to the sample. The analyst compares the color intensity of the standards to the sample and approximates the concentration of the sample. This technique has been replaced by photometric techniques; however, it is still used to determine Pt-Co color.

Photometric analysis determines the absorption of light at a specific wavelength after the addition of color forming reagent to the sample. The degree of light absorption is proportional to concentration. The absorption wavelength is a function of the chemical reaction between the color reagent and the analyte. The analyst prepares a series of standards in increasing concentration and mixes them with the color reagent. A plot of concentration versus absorbance is created. The concentration of unknown samples, reacted with the same reagent with absorbance measured under equivalent conditions as the standards, is determined by a calibration curve.

We need chemical analysis.

Testing is how we gauge whether or not treatment was effective and that our water is ok to drink or to discharge. Testing is needed to prove that water is good enough for its intended use. Our jobs, as analytical chemists, are important.

Necessary tools for use in any environmental laboratory

Balance

You need at least one top loading balance accurate to 0.1 gram for preparing reagents, etc., and an accurate analytical balance for preparing standard solutions and preparing reagents requiring small mass measurements (< 1 gram). Each balance, and especially the analytical balance, should be on a balance table kept away from drafts and influence of direct sunlight. Weighing error introduced by poor location of the balance can be detected by measuring the mass of a certified weight repeatedly at intervals throughout the day. The mass should not vary by more than 0.0001 gram if accurate work is expected. A mass difference of 0.0005 grams if preparing a stock standard of 1000 ppm from copper metal, for example, results in a 0.05 % error simply from the weighing. If, however, the analyst chooses to weigh less mass and dilute to 100 milliliters instead of 1000 milliliters, this same 0.0005-gram results in a 0.5 % error in the value of the stock solution.

Changes in weight during weighing can occur by static charge, or by differences in air density above the object being weighed. It is important that everything weighed in the balance be at room temperature. Weighing an object too hot can result in low mass measurements. Air buoyancy can introduce significant error, especially if the object being weighed is large. Try to minimize the size of the vessel being weighed. Air buoyancy and temperature effects play a role in the initial and final measurements of Total Suspended Solids (TSS) and Total Dissolved Solids (TDS). Many of the problems with TSS measurements that were related to measuring TSS within 25 milliliter porcelain crucibles are easily resolved by using larger glass fiber filters and weighing only the filter in an aluminum pan. TDS still presents a problem, especially since a very small mass of weight gain is to be measured within a much larger mass. Thus, much error in TDS measurements comes from the error in weighing the mass of the glass beaker or porcelain evaporating dish than in measuring the actual mass of the residue.

Method	Analyte
SM 2540B	Total Solids (TS)
SM 2540C	Total Dissolved Solids (TDS)
SM 2540D	Total Suspended Solids (TSS)

These are some EPA methods that make the final determination using an analytical balance. Weighing errors introduce error into your results. The smaller, or less mass, your sample container has, the better your results. For example, it is better to weigh only the filter for TSS than it is to measure the filter and a crucible. There have been plastic bags introduced that could help with the weighing error associated with TDS.

Volumetric Flasks

Volumetric flasks are second only to the analytical balance if accurate calibration standards are expected. Volumetric flasks should be high quality Class A. The calibration of Class A glassware, even purchased from a reputable supplier, should be verified. Weigh the flask empty on your top loading balance, fill it to the mark with reagent water, and weigh it again. The increase in mass in grams should be equivalent to the number of milliliters the flask is designed to hold. If not, the flask should be replaced or discarded. Volumetric glassware should be cleaned prior to use by soaking in phosphate free detergent and rinsing with tap water followed by reagent water. Store the flask upside down to prevent dust particles from falling in. Always segregate your glassware. You cannot expect to prepare accurate phosphate calibration standards, for instance, if the same volumetric was previously used to prepare a nitrate/nitrite color reagent containing 100 milliliters of phosphoric acid.

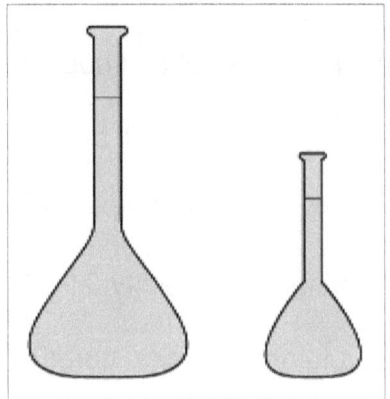

Use volumetric flasks when preparing stock standards, or reagents whose concentration affects the final result. In addition, use volumetric flasks for dilution of working standards or dilution of samples in high accuracy/precision tests. Dissolve or mix the neat material in a beaker and quantitatively transfer to the volumetric flask. Dilute to the mark with reagent water or applicable solvent, stopper, and mix by inverting. Never store solutions in the volumetric flask, instead, transfer to another container and rinse the flask. Volumetric flasks can come with ground glass fittings for ground glass stoppers. Use caution with these because they can stick if solution evaporates and there are salts between the flask and the stopper. Plastic caps and Teflon™ stoppers may be better because there are less likely to stick.

Burettes

Like volumetric flasks, burettes should be of the highest quality available. These will be used to standardize volumetric reagents and for titration methods for the determination of acidity, alkalinity, TKN, etc. Again, like volumetric flasks, burettes should be segregated for their intended use. For example, reserve burettes for sodium hydroxide reagent to be used for only sodium hydroxide reagents.

A 50-milliliter burette is sufficient for most environmental work and is capable of an accuracy of about 0.02 milliliters. In routine practice, however, the accuracy is about 0.05 milliliters. For routine work, a self-zeroing auto filling burette should be used. These burettes fill from the bottom up using a pressurized reagent bottle that forces the liquid up through the burette. Always rinse residual reagent from a previous day's work with at least one burette volume prior to any titrations. Keep the top of the burette covered with a small plastic cup to minimize evaporation and to prevent dust particles from falling in. Error from evaporation may seem small, but it does occur. Solutions prepared for burettes should be filtered and free from minute particles that could clog, or partially block, the burette tip.

Volumetric pipettes

Volumetric pipettes should be Class A and the calibration should always be verified by the user. Verify the calibration by filling to the mark with reagent water, dispensing the water into a tared beaker, and measuring the mass of water collected. Volumetric pipettes that are not accurate should be replaced or discarded. Use volumetric pipettes to accurately prepare dilute calibration solutions from concentrated stock standards. In the most accurate work, use pipettes with volumes greater than 10 milliliters and multiple sequential dilutions.

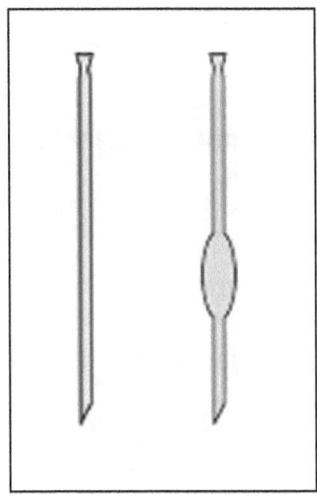

The measuring pipet is a long tube with graduations up the side. For example, a 5 milliliter measuring pipette will have 5 major markings with the space between each representing 1 milliliter and then ten smaller markings between each major mark representing 0.1 milliliter. Use these pipettes when the reagents are not the determinative reagent of the test. For example, use a measuring pipet to deliver about 1 milliliter portions of a buffer to five different sample solutions.

A volumetric pipette has a mark near the top and a bulb in the middle. These pipettes are designed to be very accurate and deliver single volumes of solution. Use these for making calibration standards and dilutions. For example, use a 10 milliliter volumetric pipette to dilute 10 milliliters of a 1000 mg/L stock standard to a 1000 milliliter volumetric flask for a final concentration of 10 mg/L. For greatest accuracy, do not use volumetric pipettes smaller than 5 milliliters.

Air displacement pipettes

These may either be fixed volume or variable volume pipettes that draw liquid into a disposable tip. The fixed volume pipettes are less complex since there are fewer moving parts. A plunger is used to both fill the pipette tip and then dispense liquid. Since the pipette tips are made of plastic it is recommended that they only be used for aqueous reagents and samples. Air displacement pipettes find common use in the environmental laboratory from preparation of calibration solutions to dilution of samples. They should be calibrated at least yearly by dispensing a known volume and weighing the mass dispensed on an analytical balance. Correction factors should be recorded on the actual pipette (using tape or a label) and applied to all measurements. For more accurate work, all dilutions using the air displacement pipette should be weighed.

Beakers and flasks

Various sizes of beakers and Erlenmeyer flasks will be needed. It's important to wash these with hot water and phosphate free soap prior to use. Rinse well with tap water and then with reagent water. As with volumetric ware, segregate the beakers and flasks to avoid cross contamination. Beakers that have contained even dilute phosphoric acid solutions, for example, can never be used to digest trace total phosphorus.

PH / ISE meters

These meters are used for adjusting the pH of reagents and buffers, adjusting the pH of samples, and detecting the end-point of acid base titrations. An Ion Selective Electrodes (ISE) is used for direct concentration (activity) measurement of ions in solution, or for end-point detection in potentiometric titrations. The type of meter should be rugged and accurate enough for extended use in the laboratory, but of minimum complexity to simplify its use. Direct reading ISE meters have the advantage of direct concentration readout (no calculations) for routine analysis.

Colorimeters

A simple filter colorimeter is sufficient for routine environmental analysis. If the laboratory wishes to expand, then a low-cost grating spectrophotometer may be a better choice, however, most common environmental tests can be done with very few wavelengths (420, 505, 540, 578, 600, 630, 660, 815, and 880 nm). The advantage of filter photometers is that the wavelength is exactly reproducible. A grating spectrophotometer wavelength can vary slightly each time adjustments from one wavelength to the next are made. Most of the inaccuracy of an environmental test is in sampling, and sample preparation making the added complexity of a high-end spectrophotometer an unnecessary expense.

Quality Assurance / Quality Control

This is a brief introduction to quality assurance of chemical measurements. Quality assurance is defined as the records kept on the results of the routine analysis of quality control samples. Many laboratories mistakenly assume that merely running quality control samples constitutes an adequate quality assurance program. This is incorrect. In fact, without proper and ongoing documentation of quality control sample results quality assurance does not even exist.

This is not a lesson in statistics; however, knowledge of statistics is required. Neither is it an analytical chemistry lesson, but without some knowledge in chemical analysis there is really no need to read further.

Quality Control

Quality Control consists of either the analysis of samples of known quantities for the purpose of verifying a method's accuracy or the repeat analysis of a sample to determine the methods precision. Quality control samples may be clean interference free matrices, or complex matrices that duplicate the sample. Results may either be recorded as absolute concentration or relative percent recovery.

Blanks consist of all reagents used in a test and may contain everything in the sample except the analyte of interest. The purpose of the blank is to assess laboratory contamination. High or variable blank values indicate a contamination that needs to be located and eliminated.

Blank Spikes are blanks to which a known amount of analyte has been added. Blank spikes determine whether significant analyte is lost during sample processing. Since the blank matrix is interference free a high blank spike result is further indication of contamination, or an inadequate calibration. A Blank Spike may also be the LCS (see below).

Blank Spike Duplicates measure the ability of a method to duplicate analytical results in an interference free matrix. Bad precision indicates either loss of analyte (lower than expected recovery) or contamination.

Matrix Spikes are real samples to which a known amount of analyte has been added. Subtracting the amount of analyte determined in an un-spiked portion enables calculation of the percent analyte recovered from samples of that matrix. The volume of spike added should be no more than 5% of the total volume of the solution spiked. The amount spiked should be about 10 times the detection limit, or 2 – 10 times the estimated un-spiked sample concentration. A spike level less than the un-spiked sample concentration will not work. Matrix spikes that fail QC acceptance criteria only apply to the particular sample spiked. If the Blank spike or LCS passes the results of that analytical batch are still valid.

Matrix Duplicates are repeat analyses of a sample matrix used to evaluate precision. If the amount of analyte is expected to be near or below the Method Detection Limit (MDL), **Matrix Spike Duplicates** are often run allowing precision to be evaluated.

Method Detection Limit (MDL) is a statistically determined number that represents the lowest concentration of analyte that can be detected with the confidence of not being a false reading. One popular calculation of MDL multiplies the standard deviation of seven replicate tests by 3.14. The replicate tests should be blank spikes with an analyte concentration 3-5 times the calculated MDL.

It is important for all users of this statistically derived MDL to realize the great inaccuracies associated with this number. The MDL that is determined by analysis of replicates made on purified water only applies to the purified water. This number generated also only applies to the analyst that made the determination and the instrument that was used. Also, statistically speaking there is no real accuracy or precision associated with this number, as variability can be as high as 100%.

Minimum Level or reporting limit is the lowest calibration standard, or a concentration of 3.18 times the MDL. The minimum level is 10 times the standard deviation of the noise and represents the point where data has an accuracy and precision of within about 30

% of its true value. A more accurate determination of the minimum level is to plot RSD and Recovery of collected multiple laboratory data and set the Minimum Reportable Level at the lowest point where both accuracy and precision are within 30%.

Calibration is a representation of a response that is in proportion to an amount. In modern instrumentation the calibration is an electronic signal relative to an amount of analyte. A graphical plot of concentration versus signal is represented by a calibration curve, which is hoped to be linear, but may be second or third order depending on the measurement method and concentration range. Calibration could, however, also represent mass measured on a balance or volume measured with a burette.

Calibration curve accuracy is often assessed using the correlation coefficient; however, a good correlation does not guarantee an accurate calibration. The slope and y intercept data should be monitored and should not change more than 10%. Keep track of calibration standard response over time.

Calibration Verification is the first step in guaranteeing the accuracy of a calibration curve. The **Initial Calibration Verification (ICV)** standard is a mid-point standard solution derived from a source other than the stock material used to prepare the calibration standards. The ICV is analyzed as an unknown and expected to be within 5% of its known value. Often the ICV standard is purchased and certified; the ICV guarantees that the calibration standards were prepared accurately. The **Continuing Calibration Verification (CCV)** is a midpoint calibration standard from the same stock used for the calibration standards. This solution is analyzed at regular intervals throughout the run to demonstrate that results are still "in control". If a CCV fails, the reason for failure should be determined and corrected. With some instruments recalibration may be necessary. Only results bracketed by acceptable CCV results should be reported.

Initial Demonstration of Capability (IDC) is a procedure used to qualify the analyst to run samples. Each analyst should perform an IDC and the documentation kept on file to demonstrate that the analyst is capable of collecting data with known accuracy and precision. At minimum, four quality control samples of a matrix similar to the sample matrix are analyzed using all steps of the procedure and compared to known acceptance criteria of accuracy and precision. Often, these Quality Control samples can be purchased.

Laboratory Control Sample (LCS) is a standard of known concentration in a matrix similar to the samples. The LCS can be purchased, prepared internally, or be a previously analyzed sample. It is preferable to have an LCS in large quantity so that the concentration is always the same. The LCS is carried through the entire sample preparation and analysis process. If an LCS fails, its quality control acceptance criteria the entire analysis batch is suspect.

Measurement techniques are moving more towards instrumentation leaving behind chemical methods such as gravimetric precipitations and volumetric titrations. A problem introduced by using instrumentation analysis is that instruments require known calibration solutions that the responses of unknowns can be compared to. As the classical volumetric and gravimetric chemical approaches to analysis and measurement are gradually forgotten we are gradually losing the ability to prepare new calibration standards for our instruments. Also, classical techniques are more accurate and precise in high purity chemical assays while instrumentation is best at trace analysis. A laboratory does itself service by maintaining classical methods and using instrumentation for trace analyses such as environmental testing, or for the verification of the purity of precipitates.

Calibration

With the exception of gravimetric (TSS and TDS) and volumetric (titrations) analysis, all environmental analyses require some form of calibration. A calibration is made by determining an instrument response from the determination of a series of standard solutions with known concentrations. The instrument response of unknowns is determined, and the concentration of the unknown is calculated by comparison to the response curve of the standards. There is error in the concentration value of the calibration standards used; however, by tight control of all mass and volumetric measurements leading to the preparation of the calibration standards the error can be minimized. Most of the error associated with calibration comes from the instrument response itself. Since there is a precision associated with the measurement, it is necessary to determine the precision of the response for each calibration concentration. Precision is determined by measuring each calibration standard in triplicate. For the most accurate calibrations a weighted fit model should be used.

Though most environmental methods prepare the calibration standards in reagent water, a matrix more closely matching the sample matrix should be used. For instance, if the matrix is seawater the calibration standards should be prepared in analyte free seawater. Many environmental methods were written to compensate for differences in matrices. Verify by preparing standards in reagent water and analyzing spiked aliquots of the sample matrix. Many environmental methods, especially some colorimetric methods, are highly pH dependent. Even if the method reagents are designed to neutralize and buffer pH preserved samples, the standards should still be prepared at a closely matching pH.

Accuracy and Precision

The accuracy and precision of a method should first be evaluated by repetitive analysis of the calibration standards. Standards at the lowest calibration level ranging to the highest calibration level should be analyzed. Run seven replicates as a minimum, and preferably more. Calculate the mean, standard deviation, and per cent relative standard deviation (RSD). Compare the accuracy (mean divided by true value) with the method. Compare the RSD with the method. They should be equivalent. Do not be too alarmed if your values are slightly worse than that of an equipment manufacturer since manufacturers typically choose the best data for publication. Once you have determined the accuracy and precision on calibration solutions, repeat the experiment with sample matrices that contain analyte at, or near, the concentration of each calibration standard. Determine the RSD of each matrix. It is not uncommon for the RSD to be higher in sample solutions. Spike each matrix with analyte at about 2 times the original analyte concentration and repeat. Determine accuracy and precision of the spiked samples. Use the spike recovery and precision data to establish preliminary acceptance criteria for future analyses.

41

Chemical Analysis of things the way they are not

With modern instrumental analysis we prepare calibrations and compare the calibration response to the response of analyte in unknown samples. The response is compared to the response of knowns and the unknown concentration is calculated. Usually calibration standards are prepared in solutions of defined matrices, often deionized water. We assume that the response of the analyte within the matrix is the same. Instrumental methods often add complex reagents, buffers, or other chemicals in an attempt to "mask" the interferences that may occur in measurement of real samples. Unfortunately, the method developer is incapable of testing every possible matrix resulting in methods that may be suitable for all the matrices tested but cannot be assumed suitable for every feasible matrix.

Many governments have established environmental regulations that require the measurement of certain pollutants. Often these regulations include maximum contaminant levels (MCL) that represent the concentration the pollutants should not exceed to protect public health or environmental health. The sheer number of samples required for testing of these pollutants necessitates a need for rapid chemical methods. Rapid chemical methods, when followed to the letter, may be suitable for the matrices they were validated with but not suitable for every matrix they get used for. The result is that governments require the measurement of certain pollutants by methods that may be seriously flawed. The laboratories and regulated entities are faced with the dilemma of following a flawed method to remain in compliance or getting the right result by not following the method and ending up out of compliance because the method has been modified.

Cyanide methods, for instance, often rely on a process called distillation that is assumed to selectively dissociate and separate metal-cyanide compounds allowing quantitative measurement of cyanide. Unfortunately, many of the components co-existing within the cyanide in the samples can create or destroy cyanide during the distillation process. The analyst cannot be confident that the final concentration of cyanide measured represents the concentration that was originally in the sample. If the interference is additive and reproducible, spiking the sample with known concentrations of cyanide does not detect the error. The only way to know if the result is in error is to analyze the sample for every potential interference, however, there are not even treatments for all of them. Another possibility is to use methods that are free from interference.

Environmental Sampling

Even if you are not responsible for collecting a sample it is important to understand the approved sampling requirements. For NPDES permits these requirements are found at 40 CFR Part 122. There are two essential "types" of samples collected "grab" or composite. Grab samples are used for pH, total phenolics, cyanide, coliform, oil & grease, and volatile organics. For all other pollutants, a 24-hour composite must be used. If the discharger only discharges 8 hours in a 24-hour period, the composite is taken from the 8-hour discharge.

The composite sample may be collected using a refrigerated automatic sampler or may be taken manually as discrete grab samples that will be composited later at the lab. Normally, samples are taken at fixed intervals, such as every hour, and composited in proportion to flow. If the flow remains constant over the entire sampling the composites may be made in equal volumes. At least 4 grab samples should be taken during the 24-hour discharge. Grab samples must be used for Oil & Grease, cyanide, phenolics, and VOC's. Each of these is collected in its own bottle requiring a person be present to collect and preserve the grab samples. It is best to collect duplicate grabs for volatiles.

Flow, temperature, and pH can be recorded continuously, however, for ease of reporting and compositing, the flow and pH measurements taken at the time of aliquot removal should be used. Preparing a worksheet for the samplers to record pH, temperature, and flow with time intervals for sampling pre-printed can be useful.

Grab samples for cyanide, phenol and VOC's are collected every 6 hours for a total of 4 grab samples. The minimum requirement for grab samples is three per 24-hour interval. If the discharge only occurs for an eight-hour period, you must still collect the minimum grab samples. I still recommend 4 grabs per sampling event.

The VOC's can be composited in the lab. While the samples are at 4C, pipet 10 milliliters from each grab sample into a 40 milliliter VOA vial for a total of 40 milliliter VOC equal volume composite. Analyze this by EPA Method 624. The cyanide and phenol grabs are run separately, and each result is reported. The same applies if Oil & Grease is collected.

Total Metals

Collect unfiltered sample and preserve to pH 2 with Nitric Acid (HNO_3). If expecting very low concentrations of metals, collect sample unpreserved and add the nitric acid at the laboratory. Digest and analyze within 6 months, except for mercury. Mercury should be analyzed with 28 days.

Dissolved Metals

Filter sample through a 0.45-micron membrane and preserve to pH 2 with Nitric Acid (HNO_3). If expecting very low concentrations of metals, collect sample unpreserved and add the nitric acid at the laboratory. Digest and analyze within 6 months, except for mercury. Mercury should be analyzed with 28 days.

Total Nitrogen (TKN), Total Phosphorus, Ammonia - Nitrogen, Nitrate + Nitrite - Nitrogen

Collect unfiltered sample and preserve to pH 2 with Sulfuric Acid (H_2SO_4) and refrigerate to ≤ 6°C. TN and TP must be digested. Analyze within 28 days. The TKN digest enables a simultaneous TKN and TP analysis. Alkaline digestions for TN can also be analyzed for TP. Ammonia and nitrate + nitrite nitrogen can be analyzed simultaneously on a suitable auto-analyzer.

Phosphate

Filter sample through a 0.45-micron membrane within 15 minutes of sample collection. Refrigerate to 4°C and analyze within 48 hours.

Phenolics

Collect unfiltered sample and preserve to pH 2 with Sulfuric Acid (H_2SO_4) and refrigerate to ≤ 6°C. Phenol must be distilled. Samples containing sulfide should be treated with copper sulfate prior to distillation. Analyze within 28 days.

Cyanide

Collect unfiltered sample and preserve to pH 10 – 11 and refrigerate to ≤ 6°C. If free or available cyanide are to be determined, store in a dark bottle. Remove chlorine with the minimum sodium thiosulfate necessary to remove the chlorine and no excess. Do not use ascorbic acid. Test for sulfide using a lead acetate test strip. If the strip darkens, dilute the sample and record dilution necessary to dilute sulfide below detection with the lead acetate paper.

Method Development

The key to method development is to start with what you know. Start with an existing method, if available, but don't assume that the developer tested all interferences, all potential matrices, or even that the reagent recipes were at optimum concentration. If the method is a standardized method, such as ASTM, EPA, ISO, or Standard Methods, verify that an inter-laboratory study was carried out and that the matrices tested were similar to the matrix you need to run. It's possible that even well tested methodology will need some minor modifications to suit your purposes.

Methods may also vary based upon instrumentation. A method may be written on one brand of instrument and require modification to operate equivalently on another brand. This is especially true with automated flow analyzers because variations in hardware not only affect the chemistry but the fluid dynamics as well.

Set the method up with the exact conditions and reagents as the approved method if possible. Then analyze your matrix along with duplicate spikes at concentrations that span the entire calibration range. If the matrix contains detectable analyte, dilute it at various dilutions. The diluted result multiplied by the dilution factor should equal the un-diluted result. If it doesn't then an interference is present. Knowing the underlying chemistry of the reaction chemistry can help you determine what the interference might be. If the result after dilution calculates to more than originally measured, the interference sequesters the signal of the analyte. Look for components in the sample that could cause your readings to be low. These could be complexing agents, etc.

If the diluted sample calculates less than the concentration that was originally measured, first verify no loss of analyte by diluting again. Look for other constituents in the sample that can also produce a signal with the chemical reaction. This could be co-precipitants in a gravimetric test, or ions that react similarly under the conditions of the test. If there are ions present that react similar, then the method is not selective, and the analyte cannot be accurately measured by that method.

For more information on method development, read *Basics of Environmental Method Development and Validation*, by William Lipps, available at https://www.amazon.com/William-Lipps/e/B06XNW1N1B?ref=dbs_p_pbk_r00_abau_000000.

Example of Applied Methodology - Oilfield water

The oil industry uses water analyses for formation identification, compatibility studies, tendency to scale or corrode, water quality, and evaluation of pollution problems or releases. The users of these data assume laboratory analyses are correct, however, data collected from multiple commercial and/or chemical company laboratories may not have the desired reliability. Lack of agreement between laboratories is understandable considering these labs are using methodology modified from methods written for fresh water. These methods were then likely modified by each individual laboratory to suit the lab's needs. Instrumentation used further complicates the issue because large dilutions are often necessary to put the analyte within range of the instrument calibration, or to dilute out interferences from the excessive salinity of the samples.

One of the first errors that can be found is simply the units of reporting. In a clean water sample, parts per million (ppm) is interchangeable with milligrams per liter (mg/L). This is true because a water sample with a low total dissolved solid (TDS) has a specific gravity very near 1.00. A brine solution, however, has a specific gravity greater than one making ppm and mg/L not equal. If ppm is reported then specific gravity should be also so that final results can be calculated in milliequivalents per liter (meq/L). It is preferred that samples are measured by volume and results be reported in mg/L and in meq/L.

Sampling oilfield waters

Water samples will be taken from a variety of sources including open pits, spills, wells, and storage tanks. These samples may be analyzed for trace organics, major and minor metals, cations and anions, physical properties such as pH and conductivity, and dissolved gases. There is no single sampling procedure that is universally applicable and sample collection techniques are determined by the analytes. Volatile organics should be collected in air tight VOA vials and immediately chilled, while samples for cations and anions can be collected in plastic containers and stored. Regardless of the analyte, if it is to be tested at a laboratory, all samples should be chilled to < 6C immediately after collection.

Conductivity, alkalinity, and pH measurements should be measured at the sampling location. Use a rugged, reliable conductivity and pH meter with sturdy factory calibrated probes. Verify calibration frequently and use the available temperature compensation features. Measure alkalinity using a titration based test kit. Do not get a simple colorimetric total alkalinity kit. Always titrate and collect both the phenolphthalein and the methyl red endpoint data. This will give you carbonate and bicarbonate concentrations that will be used later in the cation and anion balance.

Dissolved iron should be measured at the sampling location. There are reliable test kits based on the 1,10-phenanthroline colorimetric method available for this. Obtain a rugged and reliable portable photometer for this test. Do not use a kit that relies on visual estimation of concentration. Attempting to preserve samples for iron using HCL for later analysis at the laboratory will bias results high.

Dissolved gases, such as O_2 or H_2S should be analyzed at the sample location. Use either a Winkler colorimetric kit with photometer or, better yet, an optical probe for the dissolved oxygen determination. Measure traces of sulfide with AccuVac® vials and a reliable photometer.

Laboratory Analysis

Analysis of the sample for major inorganic anions and cations should proceed in a specific order once the sample reaches the laboratory. Alkalinity and pH should be determined again and as soon as the sample bottle is opened. Results should correlate with results from the field. Once these are determined conductivity should be measured next and the result written on the sample bottle. Calcium and magnesium, chloride, sulfate, sodium, and dissolved solids follow and in that order. Multiply the conductivity value by 0.6 to estimate the amount of sample needed for the TDS determination. Analytical methods chosen should measure individual analytes with little or no dilution, if possible.

Volumetric Analysis

Titration

Titration is the controlled addition of a standardized reagent solution to a solution containing an unknown mass of analyte until a desired reaction occurs. The mass of unknown analyte can be calculated from the volume and concentration of standard solution that was added. Reactions may be acid-base neutralizations, electrochemical oxidation or reduction, or the color change of an indicating compound.

Titrations require a burette, a burette stand with a holder, standardized reagent, a flask to contain the unknown sample, and a magnetic stirrer with stir bar. A squeeze bottle containing reagent water helps to rinse the sides of the titration flask or the tip of the burette aiding in the addition of small volumes, such as half a drop. The burette may be Class A glass, or a dispenser connected to a software controlled auto-titrator.

The standardized reagent should be at room temperature and transferred to the burette in a manner that prevents evaporation. An auto-filling burette that pulls solution up from a pressurized reagent bottle is preferred. The practice of pouring standard reagent into a beaker and then transferring the reagent to the burette leads to evaporation of standard reagent and can result in serious error. If an auto-dispensing pipet or auto-titrator is available these are preferred.

Measure the volume of the sample to be titrated as accurately as needed for the test. When standardizing reagents, measure the known standard solution to be titrated accurately either by weight or using a Class A volumetric pipet. Standardizations should be done in quadruplicate. A graduated cylinder may be used to measure sample; however, the cylinder should be rinsed with reagent water unless the reagent water could somehow alter the test. Orient the burette so that the solution falls directly into the sample without hitting the sides of the flask. Swirl, or stir, the solution in the flask continuously while adding titrant. Mixing should be sufficient enough for rapid dispersion while avoiding splashing or loss of sample.

Each sample should be titrated to the same intensity of indicator color, or to the same pH / mV reading. Reagent is added as quickly as possible while sample is stirring until near the end-point. As the end-point is approached, the appropriate color, or mV reading, may be seen and then go back to near where it was prior to adding reagent. This is a sign that the end-point is close but has not yet been reached.

Begin to slowly add reagent, slowing to drop by drop. Try not to "overshoot". If it looks as if one final drop is too much, suspend a half a drop on the burette tip and wash it in with reagent water.

In many titrations, the indicators are not simply present but are actual participants in the reaction. For example, iodometric titrations use starch as an indicator but the starch is not added until the titration is almost complete. Instead, a KI solution is added, and the KI reacts with the sample to create an unknown amount of I_2. The I_2 is proportional to the analyte present. I_2 is brown in color. As the titrant (often sodium thiosulfate solution) is added it reacts with the I_2 and the brown color lessens. Add starch when the sample solution becomes pale yellow. The starch reacts with the remaining I_2 and turns blue. Titrant is added until the blue color disappears. Use the volume and concentration of titrant added to determine the mass of analyte present. Dividing the mass of analyte present in the known volume of sample determines the analyte concentration in the sample.

Neutralization Reactions

An ionized molecule or compound that releases a proton, or H^+ ion, when dissolved in a solution is an acid. An ionized compound or molecule capable of accepting a H^+ ion is a base.

> **Acid + Base = Salt + Water**
> **HCl + NaOH = NaCL + H2O**
> **Low pH + High pH = Neutral Solution**

An equal amount of acid is added to an equal amount of base and you get sodium chloride dissolved in water.

We can use this neutralization reaction to measure the amount of acid or base present in a reagent or in a water sample. We slowly add a known amount of base, for instance, to a known volume of sample containing an unknown concentration of acid until the sample solution becomes neutral. By knowing the volume and concentration of base added to the known volume of sample we can calculate the amount of acid that was in the sample. This is a neutralization titration.

Pipette a known volume of sample or standard solution with an unknown acid concentration into an Erlenmeyer flask. Add a drop of indicator solution that will change color at the equivalence point. Fill a burette with standardized sodium hydroxide solution. While swirling the Erlenmeyer to mix the solution, add the standardized solution from the burette till you start to see color change and then the color dissipates. Now, slowly add the solution from the burette until a faint color persists. This is the equivalence point. Now you can calculate the concentration of the unknown acid concentration:

Volume (base added) X concentration (base) = Volume (sample) X Concentration (sample)

The equation is expressed as:

1) VC = VC

Solving for the unknown we get:

- (Volume (base) x Concentration (base)) / Volume (sample) = Concentration (sample)
- Assume the Base concentration = 0.100M
- The volume of sample added to the flask = 50 ml
- The volume of base used to neutralize the acid = 25 ml

- (25ml x 0.100M) / 50 ml = 0.05 M Acid

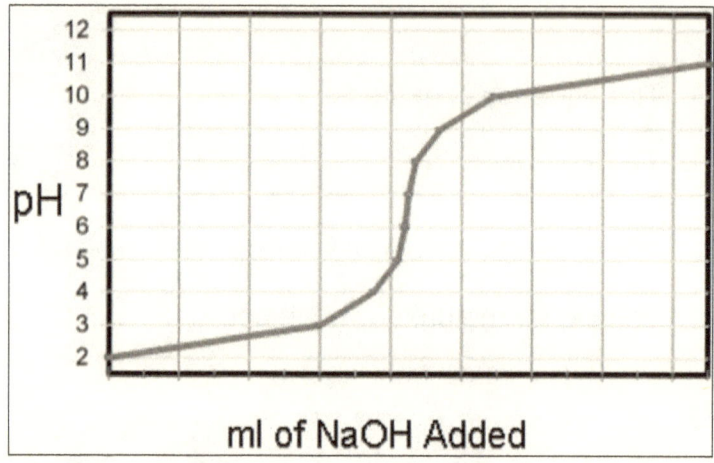

We can use a neutralization reaction to measure the amount of acid or base present in a sample. We slowly add a known amount of base to a known volume of sample containing an unknown concentration of acid until the sample solution becomes neutral. By knowing the volume and concentration of base added to the known volume of sample, we can calculate the amount of acid that was in the sample.

There is a very sharp rise in pH as the solution nears the equivalence point. In our example of a strong base neutralizing a strong acid, we can use any color indicator that changes color between a pH of about 4 to a pH of about 10. This is why phenolphthalein (about 8.3) or methyl red (about 4.5) is used in acid base titrations.

Standardization of titration reagents

The previous example could have easily been a standardization of the base solution in the burette. It is not possible to accurately weigh sodium hydroxide solutions, or to accurately measure acid solutions. All acids and bases require standardization prior to analytical work. The common practice is to standardize sodium hydroxide solutions with KHP and then standardize any acid solution by titrating it with the standardized base. Consider the previous example, but instead a known mass of KHP is added to the Erlenmeyer flask.

To standardize the NaOH, weigh 1.02 grams KHP in quadruplicate and dissolve in about 100 milliliters DI water. Add a few drops of phenolphthalein and titrate, as described above, recording the milliliters of NaOH solution used. For ease in calculation, convert the milliliters of NaOH used to liters by dividing by 1000. Then calculate the Molarity of the NaOH according to the following equation:

2) M NaOH = (grams KHP / 204.229 g KHP/M) / liters NaOH

Notice that we did not need to know the volume of the KHP solution in the flask, but only it's mass. This fact is important to know as we transition from titrations to ISE and colorimetric methods later on. For now, just realize that when titrating you are determining the actual moles (or mass) of a reactant. A titration is an absolute measurement based on chemical properties of the reactants being measured. A titration requires only accurately standardized reagents.

Ammonia Nitrogen by titration

The original ammonia and TKN methods are titrations and rely on these acid-base principles for the measurement of nitrogen concentrations. When reacted with an acid solution ammonia becomes the ammonium ion:

- $NH_3 + HCL = NH_4CL$

When dissolved is water, ammonia becomes a mixture of ammonium hydroxide:

- $NH_3 + H_2O = NH_4OH$

This NH_4OH solution is not stable and loses NH_3 gas as the solution sits. The loss of NH_3 from solution increases with higher pH, and with temperature. This fact makes it difficult to titrate ammonia directly but does make it possible to easily separate NH_3 from sample solutions by distillation. Once distilled, the ammonia molecule is condensed and captured in a standardized acid reagent. The acid reagent converts the ammonia molecule to the ammonium ion stabilizing it in solution. The remaining acid can be titrated with a standardized base and the acid consumed by the ammonia molecule captured is directly proportional to the amount of ammonia distilled from the sample.

Consider the following:

- $5\ NH_3 + 10\ HCL = 5\ NH_4CL + 5\ HCL.$

The ammonium ion in an ammonia containing sample was made basic and ammonia was distilled into an acid solution of known concentration. As long as there is less ammonia distilled into the acid than there is acid, the amount of ammonia present in the original sample can be determined by determining the amount of acid remaining. Ammonia is calculated using the difference between the original amount of acid and that remaining.

In operation a sample of known volume is added to a Kjeldahl digestion/distillation flask. The sample is made basic and immediately connected via tubing to a condenser with the exit tube immersed into a known quantity of dilute acid. The sample in the Kjeldahl flask is boiled until most of the water is evaporated and condensed inside the collection flask containing the acid. Often, a methyl red indicator is added to the collection flask enabling the analyst to verify during the distillation that the acid is not exceeded. Once the distillation is complete, the collection flask is removed and titrated with standard Sodium Hydroxide until the methyl red turns yellow. Note that the final volume of the distillate need not be known. The titration determines the absolute amount, calculated in mg, of NH_3-N distilled. Divide by the original volume, in liters, of sample to yield the mg/L NH_3-N present in the sample.

Example calculation:

- Sample volume distilled = 50 milliliters, or 0.05 liters
- Normality of standard Base = 0.02
- ppm NH_3-N = ((mls blank titration - mls sample titration)/liter sample distilled) N base x 14 mg N
- ppm NH_3-N = ((10-5)/0.05) x 0.02 x 14 = 28

Estimation of MDL:

- ppm NH_3-N = ((10-9.95)/0.05) x 0.02 x 14 = 0.28

In the first example, the volume of standard acid placed in the distillation flask (blank titration) consumed 10 milliliters of the 0.02 N NaOH reagent. The sample titration consumed 5 milliliters of NaOH. In the second example, there is very little acid consumed by the sample (0.05 ml is 1 drop). In routine analysis, the MDL by titration is about 0.3 - 1 ppm.

Alkalinity Analysis by titration

Alkalinity is an acid base titration with two endpoints. The first endpoint, at pH 8.3 is sometimes called phenolphthalein alkalinity. Titration of a water sample with an original pH higher than 8.3 to the pH 8.3 inflection point is a measure of the carbonate anion (CO_3^{-2}) present in the sample. Continue the titration to pH 4.5 measures the bicarbonate anion (HCO_3^-). The total acid used to titrate to pH 4.5 is called total alkalinity and is expressed as mg/L $CaCO_3$. It is possible that a sample with a pH higher than 8.3 could contain the hydroxide ion but this is rare. In most cases alkaline sample will only contain carbonate and bicarbonate. Using the formula available in the alkalinity methods will enable you to calculate the carbonate and bicarbonate, plus the total alkalinity. If hydroxide was present in the sample, it will be calculated also.

The pH of the endpoint can be detected with color indicators or with a pH electrode. Color indicators are faster but can be difficult to see and the ability to "see" the endpoint varies by user. The titration can be done manually by titrating from a burette, or automatically with an auto-titrator. Auto-titrators can be set to titrate to fixed endpoints, such as pH 8.3 and pH 4.5, or they can be set to determine the inflection point.

The acid solution used to titrate alkalinity should be accurately standardized with either sodium carbonate, or against a sodium hydroxide solution that has been standardized against KHP.

Other examples of titration

Other examples of titrations could include total hardness. Total hardness is titrated with an EDTA solution using Eriochrome Black T. Complex reagents should be added in wastewater analysis but are not needed for drinking water. Total hardness includes the calcium and magnesium ions. Another type of hardness is called calcium hardness that only includes the calcium ion. This is also an EDTA titration using an indicator that is specific for calcium. Hardness, reported as $CaCO_3$, is an indicator of scaling potential. Interferences can be complexed with cyanide salts, or the use of MgCDTA. MgCDTA is preferred since it is not toxic. MgCDTA also helps in seeing the endpoint. If the samples are wastewater, you may need to use the cyanide containing complex reagent.

Chloride can be titrated with either a mercury or silver salt. The mercuric nitrate titration endpoint is easier to see (diphenyl carbazide turns purple at the endpoint), however mercury is toxic resulting in a hazardous waste. The silver nitrate titration endpoint can be detected with chromate, or potentiometrically using a chloride ion selective electrode.

For more information about titrations see:

http://preparatorychemistry.com/Bishop_Titration.htm

http://www.dartmouth.edu/~chemlab/techniques/titration.html

The Ion Selective Electrode

Ion Selective Electrodes (ISE) first appeared commercially in the mid 1900's. The ISE meets a need for rapid, accurate, and low-cost analysis of frequently performed determinations such as nitrate in drinking water. The popularity of the ISE comes from its ruggedness and ease of operation, being as simple as placing it in the solution to be measured and getting an answer. It's not so simple since it must be calibrated, and since the measurements are dependent upon other ions in the solution. Even though they may be called "selective" the term should really be "sensitive" because they are not completely selective unless steps are taken to eliminate potential interferences.

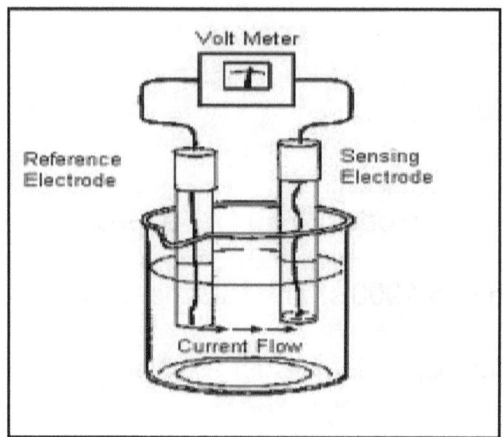

An ion selective electrode measurement requires dilute solutions and an assumption that the voltage detected between the sensing electrode and the reference electrode is proportional to the ion in solution. The reference electrode is critical to the measurement because it establishes the zero from which the voltage of the sensing electrode is measured.

A pH or ISE measures the voltage or amperage of a sensor relative to a reference electrode. The signal is determined by using a millivolt (mV) meter that can be configured to directly read pH or concentration. The most common chemical measurement made in the United States is pH with a pH meter and electrode. Other ion selective electrodes used are chloride, bromide, ammonium, ammonia, and nitrate.

An ion selective electrode measurement requires fairly dilute solutions and an assumption that the voltage detected between the ISE and a reference electrode is proportional to the concentration of ion in solution. The reference electrode is critical to the measurement as it establishes the zero from which the voltage of the sensing electrode is measured. Consider a solution with ammonium ion dissolved in it and the use of an ammonium sensing electrode. To determine concentration the response of the sensing electrode must first be determined by analysis of several solutions with a series of increasing known ammonium ion concentrations. The reference electrode establishes a zero between response from the varying ammonium ion concentrations and the sample matrix. For example, if each calibration solution contained an excess of fluoride, then a fluoride ISE could be used to establish a zero and the relative response of the ammonium electrode is proportional to the ammonium ions. The responses for the calibration standards are used to generate a calibration curve. The same excess of fluoride is added to sample solutions of unknown ammonium concentration and the response of the ammonium electrode can then be compared to the curve to estimate the ammonium concentration in the sample.

It is not practical to use a fluoride electrode on every sample, so another reference is needed. A near universal reference electrode for ISE has become the silver/silver chloride reference electrode. This electrode contains a silver wire coated with silver chloride immersed in a silver saturated potassium chloride solution. A frit keeps a potassium chloride electrolyte solution inside the reference electrode while still allowing a slow flow of electrolyte solution through the frit. The transfer of ions between the silver, silver chloride, potassium chloride, and the sample solution will establish a zero. ISE measurements using the silver-silver chloride reference can be very accurate assuming near room temperature measurement, repeatable mixing, and a constant ionic strength of the calibration solutions and the samples.

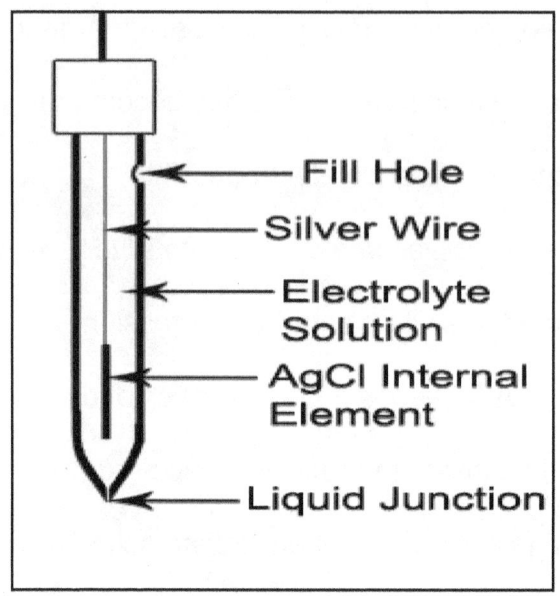

This schematic of a silver/silver chloride reference electrode shows a silver wire plated with a small amount of silver chloride immersed in an electrolyte solution. A frit keeps the electrolyte inside the probe body allowing a slow flow into the sample. The transfer of ions from inside the probe to the sample establishes a zero.

A difference between the ISE measurements and titration is that the ISE measures concentration (we assume activity = concentration). Recall that the titration determined the absolute amount in the distillation flask. Calibrate the ISE with standards of known ammonia concentrations instead of standardizing a reagent based on known chemical properties. ISE is a comparative technique. Since it is comparative, the final distilled volume must be known. This means that once a sample is distilled and collected into the absorber solution, the ammonium ion concentration cannot be determined until the solution is first brought to a known volume.

Carry the distillation out as described in titration above. Remove the receiving flask and dilute the solution in it to a known volume. Then take an aliquot of the distilled sample solution, add an ionic strength adjuster, and measure ammonium by the

Routine operation of ISE direct potentiometry

One of the advantages of ISE analysis is that the equipment needed to do an electrode analysis is not very expensive. The prime requirement is a voltmeter capable of reading +/- 0.2 mV or better. Most pH meter manufacturers will manufacture a combination pH/ISE meter with built in automatic calibration allowing direct concentration readings. Another advantage is the combination electrode that combines the indicator electrode and the reference electrode in a single unit. These electrodes are about the size of a felt tipped marker.

For accurate ISE measurements, all solutions should be stirred. The stirring speed and the immersion depth of the electrode in each solution should be kept constant. Samples can be poured into beakers, or small plastic cups. Use a high-quality stirrer that does not generate heat, and magnets with an inert coating that can be cleaned between samples.

Calibration is necessary in direct potentiometric analysis. Calibration is very simple, but attention to detail is still necessary. Prepare a series of known concentrations and pipet equal amounts to individual beakers, or plastic cups. Add a magnetic stir bar, and then add the appropriate amount of ionic strength adjusting solution. This solution varies with electrode/analyte and its recipe will be provided in the method or by the electrode manufacturer. Follow the manufacturer's instructions for direct concentration reading or record the mV of each standard solution and plot versus the log of the concentration. Stir at a constant rate and always immerse the probe to the same depth. Measure samples under the same conditions as standards. Then either record the direct concentration reading or calculate the concentration from the calibration curve.

Determination of Ammonia by ISE direct potentiometry

The sample is distilled as described in determination of ammonia by neutralization titration. Once distilled, the ammonia exists as the ammonium ion in an unknown volume of distillate. This solution must be brought to a known volume prior to analysis. Although there are ammonium ion selective electrodes, we recommend the use of an ammonia gas selective electrode. The gas sensitive electrode does not suffer from interferences from the sodium ion and can be used in determinations of ammonia and TKN in non-distilled samples. (If samples are not distilled, you must be able to demonstrate that distillation is not necessary). Once the distillate is adjusted to a known volume, pipet enough sample into a plastic cup so that it can be stirred evenly, and the electrode immersed without being hit by the magnet. Begin stirring the sample and then add the ionic strength adjuster solution (a strong base reagent that converts ammonium to ammonia). The ammonia passes through a membrane and a response proportional to the ammonia concentration is measured. Calculate the concentration of ammonia in the original sample:

- Mg/L NH_3-N by ISE x (mls diluted to /mls distilled)

The gas sensitive electrode selectively diffuses ammonia through a semi porous hydrophobic membrane into an electrolyte solution held by the membrane between the electrolyte solution and the sample. The reference electrode is an integral part of the electrode, and the indicator electrode measures pH changes of the electrolyte. The membrane needs to be changed about once every two months depending on the sample. Response times can be sluggish for low concentrations of ammonia, and slow to recover after high concentrations. If measuring distillates, it is only necessary to use base solutions as an ionic strength adjuster, however, if direct determinations of ammonia, or TKN digests, are attempted without prior distillation then complex reagents need to be added as well to prevent fouling of the membrane. The diffusion process is highly dependent upon the temperature and stirring rate of the solution; it is necessary to measure all standards and samples at the same temperature while stirring at a reproducible speed. Note that stirring should be fast enough to ensure adequate mixing, but slow enough not to produce a visible vortex.

For more information on the direct ISE analysis

of Ammonia see:

http://www.coleparmer.com/TechLibraryArticle/971

Fluoride Analysis by ISE

The fluoride ion selective electrode has almost 100% selectivity for the fluoride ion. The only interference, hydroxide (OH^-) is easily controlled by buffering the sample and standard solutions to pH 5.5. As with all ISE measurements, the fluoride electrode is measuring activity instead of concentration. In the dilute solutions (0 – 10 ppm) that we are measuring in drinking water samples activity is equal to concentration.

The Fluoride ISE consists of a lanthanum fluoride crystal and uses a silver /silver chloride reference electrode. The reference electrode is internal, and the electrolyte solution contains fluoride. There is a potential between the silver/silver chloride wire and the internal electrolyte, and a potential between the internal electrolyte and the internal surface of the lanthanum fluoride crystal. This potential establishes a zero. Changes in the fluoride concentration touching the outer surface of the lanthanum fluoride crystal produce a millivolt reading that is proportional to concentration.

Although the fluoride electrode is highly specific for fluoride, it cannot detect fluoride that is bound to metal ions, such as Al^{+3} or Fe^{+3}, in the solution. The Total Ionic Strength Buffer, known as TISAB, buffers the sample and standard solutions to pH 5.5 and contains complexing reagents that free bound fluoride. Low spike results are usually due to presence of metal ions in excess of the TISAB's complexing capacity. Samples with low spike recoveries may require distillation.

As with all monovalent ISE measurements, the

slope should increase or decrease by about 57 mV per each decade change in standard concentration. Calibration should always encompass at least one decade rise in concentration and even with direct reading ISE meters you should verify the millivolt readings with each calibration. Always use plastic cups or beakers with Teflon coated stir bars. Avoid glass to the extent possible. Stir samples and standards at a constant rate, make sure standards and samples are all at room temperature, and always immerse the electrode to the same depth for each measurement. Be sure to thoroughly rinse the electrode with deionized water between measurements, and blot dry before inserting into the next solution.

Store the electrode in a high concentration, such as 100 ppm F, standard containing the proper amount of TISAB. Even though the lanthanum fluoride crystal is very insoluble in water, it is possible for some dissolution to occur if it is stored in reagent water. The crystal can be cleaned using the cleaning cloth supplied with the electrode. Be sure to follow the manufacturer's instructions and be careful to not scratch the crystal.

Nitrate by Ion Selective Electrode

The nitrate ion selective electrode is not as selective as the fluoride electrode, or the ammonia electrode, however, it can be used for accurate determination of nitrate ion in fairly clean matrices such as drinking water and some waste-waters. The nitrate electrode should not be used in seawater, brackish water, or samples that may contain organics and surfactants. An advantage of the nitrate electrode is ease of use, and that it directly measures nitrate as opposed to the reduction colorimetric methods that measure nitrate plus nitrite combined.

The nitrate ion selective electrode can be used in samples with a pH range between 2 and 11.

Since nitrate is a monovalent anion expect about a 57-millivolt difference between standard solutions that differ in concentration by a factor of 10. Always calibrate with standards at least one order of magnitude apart, and always verify the change in millivolt readings (this verifies the slope). The slope decreases as the electrode membrane fouls and becomes less sensitive.

The nitrate electrode consists of an internal solution, usually ammonium sulfate, containing nitrate that is in constant contact with an electrode. The solution is separated from the sample by a thin membrane that selectively allows nitrate to pass through. The double junction reference electrode, either as a separate probe, or as an integral part of a combination electrode establishes a zero. When the electrode is immersed in samples any nitrate present diffuses through the membrane changing the potential of the internal electrode. This potential change is proportional to nitrate concentration.

Any ion that can diffuse through the membrane is an interference. Interferences are usually monovalent ions such as perchlorate, iodide, and high concentrations of chloride. Since the electrode works by diffusion of nitrate through a membrane, it is highly dependent upon temperature. Ensure that all calibration standards and samples are the same temperature. Always stir samples and standards at a constant rate, and always immerse the electrode to the same depth. Anything that can react with the membrane interferes. Surfactants and organics can coat the surface of the membrane preventing ions from diffusing through. Some organics could actually dissolve the membrane.

Store the electrode by soaking in a 100 ppm Nitrate standard if it will be used within the next few weeks. For longer term storage follow the manufacturer's instructions.

Colorimetry

A drawback of direct potentiometry, or ISE, is the slow response times and lack of sensitivity at lower concentrations. As we need to measure lower concentrations of nutrients, we have come to rely on colorimetric methods of determination. While visible colorimetry came in to use in the late 1800's, we will devote our efforts to electro photometry. Photoelectric colorimeters came in to use in the mid 1900's. Analysis with a photometric colorimeter is also a comparative technique requiring calibration with known standards. The analyte is reacted with reagents under controlled conditions that produce a colored compound, called a chromophore that absorbs light at a known wavelength. The amount of light absorbed is proportional to the analyte concentration. Measurements are made in cuvettes of fixed path length. Calibration of the colorimeter takes advantage of a concept known as Beer's law.

Explanation of Beer's Law

It is a well-known fact that the more beer you drink the drunker you get. This applies not only to beer, but to any alcohol. The more accurate statement is the more alcohol you consume the more intoxicated you get. Thus, Beer's Law can be expressed as:

- A = abc, where:
    - A = amount of intoxication
    - a = alcohol type (beer, wine, liquor, etc)
    - b = bottle size
    - c = cups consumed

Suppose someone is drinking 5.0 beers, in pint size mugs or cups. Their intoxication level increases proportionally with the number of cups consumed. By measuring their intoxication level you can estimate the number of cups consumed. This holds true as long as the alcohol type and the cup size are held constant.

The rate of intoxication (slope of the line) is a function of the alcohol type and the bottle size, or of ab. As long as ab is kept constant the number of cups consumed can be estimated.

To increase the rate of intoxication (slope) we must increase the type of alcohol (a) or the bottle size (b).

Of course, we know that Beer's law is not really about alcohol consumption. Beer's law is a measurement of Absorbance in relation to molecular absorptivity, path length, and concentration. But the same principles apply. In using Beer's law to estimate concentration, you measure the Absorbance of known standards reacted under controlled conditions and measured in the same, or equivalent length, cell. Beer's law is expressed as:

- $A = abc$ where,
    - A = Absorbance
    - a = molecular absorptivity
    - b = path length
    - c = concentration

The reagents used, their preparation, and the amount added to a fixed volume of standard or unknown solution determines the absorptivity of a reacted sample. When recipes are followed exactly with pure, fresh reagent the absorptivity is a constant. The path length is fixed by the walls of the cuvettes. The only variable is concentration.

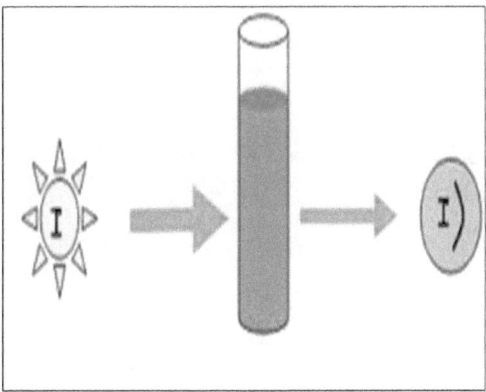

Beer's law is a measurement of absorbance (A) in relation to the molecular absorbtivity (a), path length (b), and concentration (c), (A = abc). To determine sample concentration, you measure absorbance of known standards in a cell with a fixed path length, reacted with reagents under controlled conditions. Then you plot a calibration curve of concentration versus absorbance. Next, you react sample solutions and measure absorbance using the same reagents and conditions used to prepare the calibration curve. Calculate the sample concentration using absorbance of the sample.

If you were to develop the color and measure it with the naked eye, then you would be measuring the increasing intensity of the developed color as proportional to concentration. You would make a series of standards and add reagent. You get some sample add reagent, and then compare color. The darker the color the more analyte there is in solution. This is exactly <u>not</u> what you are measuring when doing Absorbance, in Absorbance measurements you are measuring loss of color. Thus, the darker it gets of one color; the dimmer it gets of the color you are measuring.

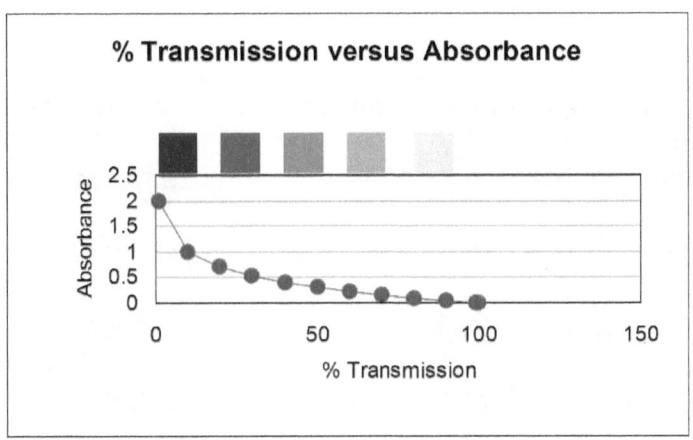

The spectrometer measures (as Absorbance) the absense of light hitting the photometer. As the color looks darker to us, the light hitting the detector becomes less.

It is important to remember that in Absorbance measurements you are measuring an absence of light. This means that as concentrations get lower, the detector must be able to distinguish what is missing and not measure what is there; it is easy to see the absence of a lot of something, but much harder to notice a few things missing from a lot of things.

Imagine a jar containing 100 marbles. These 100 marbles represent 100% Transmission because there was no chromophore present to absorb any of the light passing through. Now remove 1 marble to represent a 99% Transmission, or 0.0044 Absorbance, and a detection of the chromophore. Assume the jar represents a photometric detector. The zero Absorbance reading, or 100 %T, is the 100 marbles, and the 99 dots represent a signal. This should help explain why it is difficult for the photometer to distinguish very small amounts of analyte; the photometer needs to detect very small amounts of photons missing.

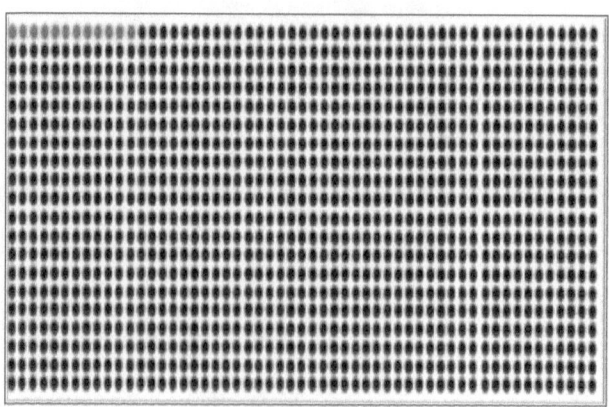

This rectangle contains 1000 dots, representing 100% Transmission. There are 12 red dots representing just over 99% transmission, or 0.0044 Absorbance. Assume this rectangle represents the entire surface of a photometric detector. The zero Absorbance reading would be if all these dots were black, and the 0.0044 absorbance is represented by removing the red dots. Of course, in reality, the missing dots are scattered randomly (shown here in order so that you may more easily see them). This graphic should help to explain why it is difficult for a photometer to detect very small amounts of analyte.

Photometric analysis of Ammonia

There are several chromophores that may be used for the measurement of ammonium ion, or ammonia. Originally, most ammonia analyses used the Nessler reagent. The Nessler reagent contains mercury and so its use is diminishing. We will not discuss it further. Colorimetric methods for ammonia analysis that do not use the Nessler reagent are usually some form of the Berthelot reaction. The Berthelot reaction combines ammonia with chlorine to form chloramine. The chloramine then reacts with phenol in strong base to form a blue colored dye called indophenol. The reaction can be catalyzed with nitroferricyanide and heat. The indophenol color is a blue or bluish green and Absorbance is measured at 630 - 660 nm. (630-660nm is red light, and reaction looks blue; we are measuring the absence of red).

The indophenol reaction is very sensitive allowing detection limits in the low parts per billion. Precursors for the dye are usually phenol or salicylic acid, but other phenolic compounds will also work. Salicylate is gaining popularity because it is not as toxic, nor does it smell as bad, as phenol. The phenol reaction is slightly more sensitive.

Since the reaction takes place at high pH, complex reagents must be added to un-distilled samples. Currently, however, the EPA requires that samples be distilled. Interestingly, even though the EPA requires that samples be distilled the methods still require complex reagents. Care should be taken in choice of complex reagent. The most popular one, EDTA, actually interferes with the analysis.

Ammonia Analysis by Manual Spectrophotometer

Prepare a series of calibration standard solutions in concentrations that will produce Absorbance readings no more than about 0.800. Prepare the lowest standard at, or near the Minimum Reporting Level, or that will read about 0.01 -0.05 Absorbance. Prepare at least five standards and a blank. The matrix should duplicate the sample matrix in acid concentration; if samples are preserved or distilled into dilute acid the standards should be prepared in dilute acid.

The manual phenate method can be found in Standard Methods for the Examination of Water and Wastewater-, and is summarized as follows:

1. Dissolve 11 grams phenol in 100 milliliters ethanol.

2. Dissolve 0.5 grams of sodium nitroferricyanide in 100 milliliters of water
3. Make a stock citrate buffer by dissolving 200 grams of sodium citrate and 10 grams of sodium hydroxide in 1000 milliliters of water.
4. Prepare an oxidizing solution by adding 25 milliliters of Clorox to 100 milliliters of the citrate buffer.

Color development of ammonia samples by the manual phenate method:

1. Pipette 25 milliliters of blank, standard, and sample into separate 50 milliliter volumetric flasks.
2. Pipette 1 milliliter of the phenol reagent to each and mix.

3. Pipette 1 milliliter of the nitroferricyanide reagent to each and mix.
4. Pipette 2.5 milliliters of the diluted Clorox-citrate reagent and mix.
5. Stopper each flask; place them in a dark place at room temperature for a minimum of one hour.
6. Read the Absorbance in a spectrometer at 640nm.
7. Prepare a calibration curve and calculate the concentration of ammonia nitrogen in the samples by comparing to the curve.

For more Information on the manual phenol methods see:

http://ceeserver.cee.cornell.edu/mw24/cee453/NRP/ammonia%20phenate.htm

http://www.standardmethods.org/store/

Nitrite Nitrogen by Manual Colorimetry

Nitrite nitrogen measurement by manual colorimetry is fast and simple. The color reagent is very specific for nitrite resulting in little interference.

Nitrate plus Nitrite by manual cadmium reduction and manual colorimetry

The manual method for nitrate plus nitrite uses the same colorimetric reagent as used for nitrite requiring a preliminary reduction of nitrate ion to nitrite ion. This reduction step is done by passing a buffered sample solution through a column containing cadmium metal granules. As the nitrate ion travels through the cadmium bed it is reduced to nitrite. Once reduced, the sum of the nitrate reduced, and the nitrite originally present in the sample are measured as a single nitrite nitrogen result. The nitrate nitrogen result is obtained by subtracting nitrite nitrogen from the nitrate plus nitrite value.

A cadmium reduction column is prepared by packing "copperized" cadmium granules in a glass tube through which buffered sample solution will be poured and allowed to gravity percolate through. Cadmium is copperized using a copper sulfate solution as follows:

- Add enough cadmium granules to pack the column to a beaker (about 25 grams).
- Rinse with 1+1 HCL
- Rinse with buffer reagent
- Swirl with 100 milliliters of 2% Copper Sulfate Solution until blue color fades
- You should see a brown colloidal copper

precipitant

- Rinse with buffer reagent till all copper is washed away

Fill column with buffer reagent and slowly add the granules. Never let the column go dry. Activate the column by passing at least 100 milliliters of 1 ppm nitrate standard through it.

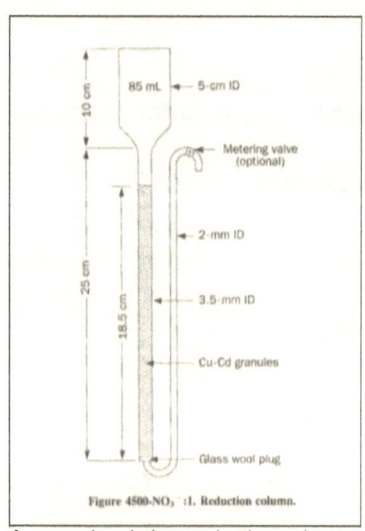

A manual cadmium reduction column

Residual Chlorine by Manual Colorimetry

There are two DPD methods for residual chloride by manual colorimetry. One method is total residual chlorine and the other measures free chlorine. The only difference between these two methods it whether or not potassium iodide was added. The total residual chlorine method adds potassium iodide to reduce all chlorine, either free or bound to ammonia, to iodine and then measures the iodine produced. The absorbance is proportional to chlorine originally present in the sample. The free chlorine method reacts the DPD directly with the sample and does not detect chlorine bound to ammonia. The difference between the two results is used to estimate the concentration of chloramines, however, chloramines result in a small false positive (0.1 – 0.3 mg/L) in the free chlorine test.

Oxidizers in the sample, such as

permanganate can interfere as well as turbidity. In drinking water permanganate may be present, however, turbidity should not be. In wastewater effluents where low chlorine residuals are often measured you should take care to either filter the sample, or zero each measurement with sample to which color reagent has not been added. Let's examine the filtration statement a bit. Most wastewater methods/regulations distinguish between total and dissolved by filtration of the sample through a 0.45-micron filter. By this definition it is impossible to measure total residual chlorine on a filtered sample. However, remember that total residual chlorine is a measure of free chlorine and chlorine that is bound to ammonia (chloramines). Both free chlorine and chloramines are water soluble, will not be bound to particulate matter, and will pass uninhibited through a 0.45-micron filter. If you filter a sample for total residual chlorine you will get the same result as a non-filtered sample minus the turbidity.

Colorimetric Test Kits and Pre-packaged reagents

Most of the simple colorimetric tests we use in environmental chemistry are for analytes that are not very stable once dissolved in solution. Also, the reagents, once dissolved, are not very stable either. The short holding time of the analyte combined with the short shelf life of standard and reagent solutions has led to a proliferation of test kits including prepackaged reagents and photometers with pre-stored calibration curves. As we discussed, the only variable when path length is kept constant and reagents are fresh is concentration. Therefore, it is scientifically valid to use an instrument stored calibration as long as reagents are consistent. Verify the calibration with a series of check standards that span the range you are interested in.

Automated methods of analysis

We have looked at titration, ISE, and manual Colorimetry. All of these methods are sufficient for the tasks they were intended to do. If lower concentrations need to be measured, or if there are too many samples making it impossible to keep up with the workload automation is needed. You can automate titration, and you can automate ISE, however, even with automation these techniques cannot achieve the throughput, nor the sensitivity, of a continuous flow colorimetric analyzer.

I began in the laboratory testing industry in 1979 at a 7 – 10-person laboratory with very little automation. We had no auto samplers, or no continuous flow analyzers. Computerized data acquisition didn't really exist yet. Over the years we slowly added some computerized integration, but we were always scared to risk the expense of acquiring full automation. It wasn't until 1986 that I experienced the power that can be obtained simply by the addition of an automated segmented flow analyzer. The benefits far outweigh the minimal cost. The increased throughput so improves your laboratory's capacity to run samples that you will kick yourself for doing without it. You can easily expand from tens of samples per day to hundreds per day. You can confidently bid on contracts you were always hesitant on before. The quality of the data improves, detection limits go lower, and employees are much, much happier.

What is automation?

Automation is having a machine do things for you releasing you from repetitive tasks and allowing your laboratory to analyze more samples per day and at the same time, improve long term method performance.

Anyone looking to increase throughput and increase overall laboratory capacity should consider some form of automation. Adding automation to your lab allows your employees to spend their valuable time doing things other than those same old, same old tasks. Automating repetitive tasks lowers systematic errors that result from boredom and operator fatigue. Automation also ensures that analyses are done exactly the same way over and over again ensuring quality control criteria is always met. Extra benefits include waste minimization and decreased costs. If you are looking to increase your labs efficiency, then you should consider automation.

An automated analyzer works while you are at home. Great profit can be realized by this unattended nighttime operation. It can be considered equivalent to having another shift. Day-to-day reproducibility is increased because regardless of the operator the analyzer does all chemistries exactly the same way every time. Reagent use is less, so hazardous waste generation is less saving you time and hassle with waste disposal.

It's all about running more samples and this benefit from automation is immediately realized. Efficiency improves giving you more time, and the added time enables you to run more samples. Your staff is happy because they can do something useful rather than stand around pipetting or reading colorimeters and Ion Selective Electrodes.

More samples per day results in more profit. Even if you are a research, or water treatment facility that does not charge per test you still profit by letting the instrument do the work for you. An automated instrument lets you gather more information and keeps working while you're at home sleeping. It is feasible to accomplish three times the work than you could do manually in an 8-hour day.

The simplest form of automation is the magnetic stirrer. Automating just makes life easier. Imagine using a stir rod for hours to get something to dissolve.

Other examples are auto filling burettes, vacuum filtration, bottle top dispensers, etc.

Imagine using a swing balance instead of an electronic one and waiting for the balance to stabilize just for a weight. Recall that all peaks were once measured by cut and weigh, or by heights measured individually with a ruler.

To a large degree, we have become spoiled in our expectations to allow the electronics do everything for us.

Almost any method that can be done manually can be automated. Time consuming steps such as manual titrations can be replaced using instrumentation. Instruments can do digestions, distillations, dilutions, and filtrations all for you. More importantly, is that these methods are always duplicated <u>exactly</u> ensuring that quality control procedures are met. Let me clarify that when I say <u>exactly,</u> I am referring to the variability you get between different analysts when they run the same thing, or the single analyst variability that occurs with fatigue accompanying the same analyst doing the same thing over and over again. The automated instrument always does the same thing regardless of who turns it on. It works day and night without getting tired.

With an instrument you can automate labor intensive extractions and digestions. In-line sample preparation procedures include digestion to replace batch heating blocks, distillation to replace macro and midi glassware, solvent extraction to replace separatory funnels and significantly reduce organic solvent consumption and waste, gas diffusion such as that used in the cyanide method to remove the analyte from a potentially interfering matrix and liquid dialysis to remove particulates and turbidity from the sample matrix. The analytes to which the various in-line sample preparation procedures apply include digestions and distillation for cyanide, distillation for phenol, UV digestion for total nitrogen and phosphorus and solvent extraction for surfactants.

Almost any method that can be done manually can be automated. Regardless of the automation technique they all do the same thing. And that is making your life easier.

Manual methods are by nature error prone because analysts get tired, or because pipetting, mixing, and so forth just varies between analysts. Manual methods limit a laboratory's ability to do more samples per day. An analyst must stand at the machine to record every measurement. It is much better to have that analyst performing other tests, or doing other things, while the automated analyzer is happily measuring samples for you.

An automated method, or analyzer, should be easy to use and easy to understand. If it takes all day just to get it going, then what is the point? An automated analysis should give you better long-term precision and should significantly lower labor and reagent costs per test.

Should you automate?

The number one question on deciding whether to automate should always be how many samples do you, or will you, have followed immediately by how many different tests will you be doing per sample? The answer to these questions helps to determine the most profitable way (in time saved and money earned) to automate your laboratory.

Now that you have determined how many samples and how many different tests you have, or will have, the next questions are whether there are a lot of tests on the same sample or a lot of samples for the same tests. The answer to these questions is critical.

The answers are critical because the best mode of automation is depending on the way your laboratory operates, and its sample load. Often, laboratories experience both of these situations. For example, a laboratory may receive hundreds of ammonia samples per week making a low-cost flow analyzer economically feasible. The same lab may also be receiving occasional phosphate and nitrite samples that if analyzed on the same flow instrument would cause lost time in the ammonia analyses. In this case, a high throughput non selective analyzer for the ammonia plus a lower throughput selective (or discrete) analyzer for the short holding time phosphate and nitrites would serve to not pressure the ammonia analyst and ensure that the phosphate and nitrite got done. In this example, the selective analyzer could actually be used at, or very close to, sample log in making high pressure "rush" samples (or those that come in near holding time expiration) a snap because they are easily analyzed by the person who knows that they are there – the login person puts them on.

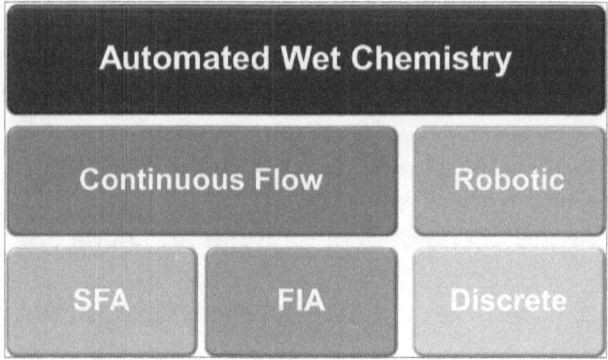

Two available analytical techniques capable of automating wet chemistry analysis are continuous flow analyzers (CFA) and robotic analyzers, also called discrete analyzers. CFA may be further subdivided into segmented flow analysis (SFA) and flow injection analysis (FIA). Continuous flow analyzers transport reagents and sample continuously through tubing. Discrete analyzers deliver reagents and sample to individual reaction vessels.

Continuous Flow Analyzers

A continuous flow analyzer consists of a pump that moves reagents through a manifold composed of tubing with tees for the addition of reagents and sample and coils for mixing sample plus reagent and creating time delays that allow reactions to proceed. Within the manifold there may be a series of modules added that can digest, distill, filter, or perform sample processing. How the sample volume is measured and injected into the tubing is a primary differentiator between SFA and FIA, the two major divisions in continuous flow analysis. SFA uses air segmentation to separate sample segments allowing for better mixing and decreasing carryover between sample segments. FIA does not segment with air and relies on the dispersion that occurs within the tubing to mix the sample. Without air segmentation, FIA chemistries must proceed rapidly through the cartridge.

History of laboratory automation of wet chemical analysis

The first auto-analyzer was introduced by the Technicon Corporation in 1957. This analyzer automated wet chemical reactions by mixing reagents and sample in a continuous stream flowing through small diameters of tubing. The reacted stream passes through a photometric detector. When there is only reagent and no sample passing through the detector a recorder only measures baseline. When sample passes through the recorder measures the Absorbance as a peak rising above the baseline. A sample analysis, called the peak trace, is a series of peaks of various sizes with the size of the peak being proportional to concentration.

The original, commercial, continuous flow analyzer segmented the stream of reagents with air, or an inert gas, preventing dispersion as the sample and reagent mixture travelled down the tubing. Segmenting the flow is called Segmented Flow Analysis, or SFA. Another similar technique does not segment the flow and is called Flow Injection Analysis, or FIA. Many believe that SFA came before FIA, and commercially this is so, but it is only logical to assume that the inventor (Leonard Skeggs) of SFA did not intuitively try segmentation first. It is most likely that Skeggs discovered the addition of segmentation to minimize dispersion by accident.

As Skeggs was experimenting with running reagents through tubing he was limited to the tubing available to him in the early 1950's. Teflon® tubing, if even available would not have been readily available to Skeggs. The internal tubing diameters necessary for FIA (0.5 – 0.8 mm) were not available either. What was available was glass. The first auto-analyzer used glass tubing and glass coils. I believe that as Skeggs was experimenting with pumping liquids through his machine he accidentally discovered that introducing air segments made the peaks sharper. Regardless, I believe that at the time, the early to late 1950's, what we know as FIA was not possible, virtually forcing the first commercial automated chemistry analyzer to be a segmented flow analyzer. It is the developments that came about as a result of the acceptance of continuous flow that prepared the path for eliminating segments. These developments include plastic tubing, narrower bore glass tubing and coils, heaters, dialysis membranes, and many other items that we consider commonplace today.

While Segmented Flow Analysis came first, and paved the way for Flow Injection Analysis, we are going to examine FIA first. FIA is simply a subset of SFA.

Flow Injection Analysis

In flow injection analysis a known volume of sample aliquot is injected into a continuous flowing stream of reagents. As the sample and reagent flow through the tubing they mix, react, and form a measurable product. This reaction product flows through flow-cell that detects and records a continuous signal. When only reagents are passing through there is baseline. When sample and reagents pass through there are peaks.

The interaction of the sample and reagent with the inner walls of the tubing forces them to mix. Kinks and turns in the tubing minimize dispersion. A reaction product will form. The peak shape is a result of the extent of reaction and the amount of dispersion that happens as the sample is passing through. The longer the sample is reacting with reagent, the greater the response, however, the longer the sample is in the tubing the higher the dispersion. Thus, there's a tradeoff. The length of tubing determines the reaction time (response) but also determines the amount of dispersion (dilution); the higher the dispersion the lower the throughput and also the response. As a general rule, FIA methods should inject, react, and measure the sample all within one minute or less.

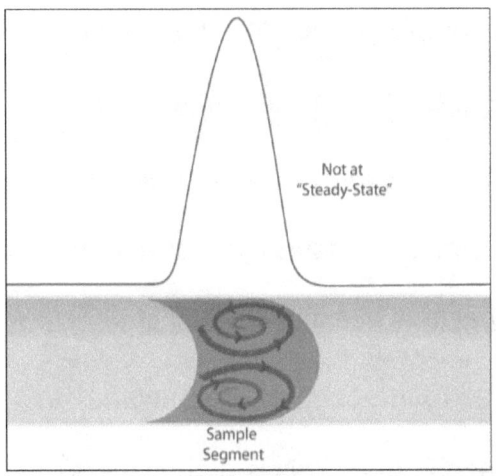

FIA peaks are Gaussian shaped as a result of dispersion. FIA methods control dispersion to obtain adequate mixing and peak response. An ideal peak is pictured here. Notice that there is no flat top, FIA methods do not attain steady state, meaning that all reagent has not completely reacted with the sample. FIA methods are designed to get sample in and out of the cartridge in a minute or less.

If the sample passes through too quickly you get a leading peak (looks like a shark fin) and low dispersion. This is adequate if there are no reactions and you are simply measuring a parameter such as pH. If the sample is in the analysis too long, the peaks become broad to the point where they may even merge with each other. This is high dispersion. A reaction that requires so much time as to become almost immeasurable by FIA should be measured by SFA.

Medium dispersion is ideal for a FIA reaction. These peaks are near Gaussian in shape. There is sufficient washout between sample injections so that peaks fall completely to baseline before the next peak passes through the detector.

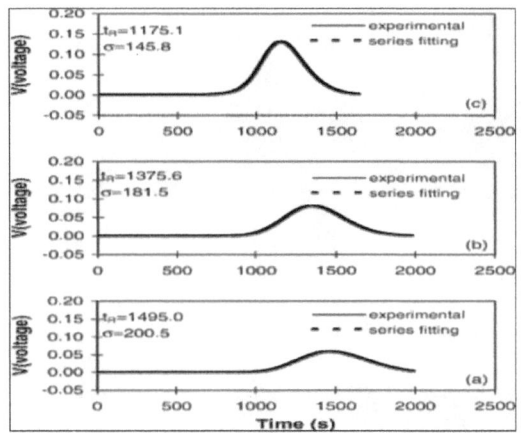

If the sample is in the FIA cartridge too long (bottom), the peaks become broad to the point to where they may merge with each other.

Remember that the peak shape is a function of the time it takes for a sample to travel from injection to the detector. It is also a function of the straight paths of tubing, or the coils, bumps, and turns. The straighter the paths, the higher the dispersion. Dispersion is also a function of internal tubing diameter and flow rate. As you can see, modification of a FIA analysis cartridge can significantly alter sensitivity and throughput.

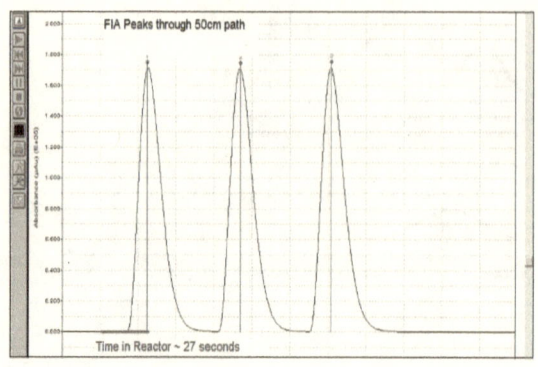

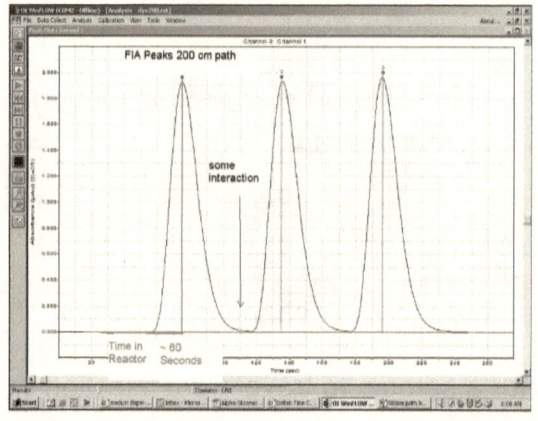

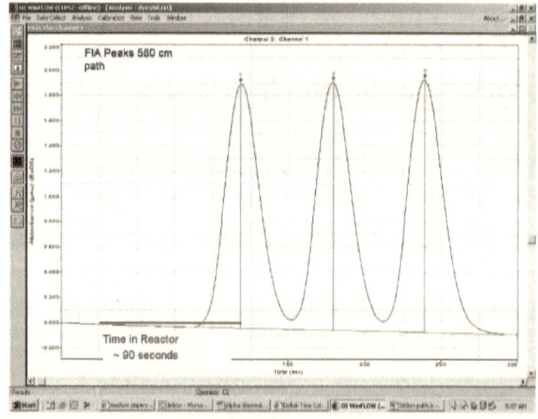

The longer the FIA path, the broader the peaks get. Eventually they may merge together. For longer reactions, use SFA.

Segmented Flow Analysis

The purpose of the air injection in segmented flow analysis is to minimize dispersion. Minimizing dispersion allows slower reacting chemistries to be reacted for sufficient time (time = length of tubing) without excessive peak broadening.

A true SFA peak should look rectangular. The width is proportional to the time the sample probe pulled in sample (amount of sample injected) and the height is proportional to concentration. Now, overlay a bell-shaped curve to represent a perfect FIA peak of the exact same injection volume. The height of the FIA peak is lower, and the width is wider. In true SFA, peaks are rectangular with no dispersion. Since FIA relies on dispersion to mix, FIA peaks are the rectangle smoothed out into a bell curve. Because there is theoretically no dispersion with SFA, SFA is ideal for chemistries with longer reaction times; the segments prevent excessive peak broadening as sample passes through the tubing.

A rule to remember - if the chemistry does not require long reaction times then the choice between FIA or SFA does not matter. Simplest is always the best. Choose FIA for the simpler chemistries with fast reaction times. When reaction times are long, choose SFA.

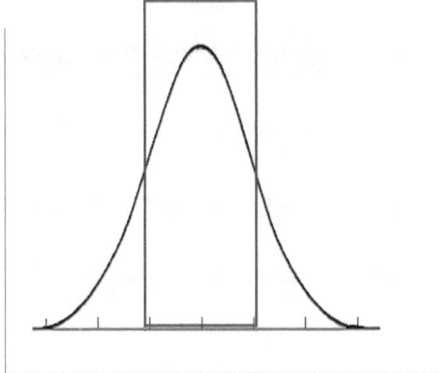

The purpose of air injection is to minimize dispersion. A theoretical SFA peak is a rectangle with a width proportional to the time the sample probe pulled sample, and a height proportional to concentration; there is no dispersion in theoretical SFA. FIA however, relies on dispersion. Therefore, a FIA peak is always broader with a lower peak height.

If the peaks are in the system for a long time, FIA peaks will be broad merging into one another, while SFA with identical sample volumes and wash times result in large separations between peaks. In these instances, where long reaction times are necessary, SFA is capable of much higher throughput than FIA.

For example, in a long reaction time method, SFA peaks are narrower, resulting in more peaks in the same amount of time. Note that throughput, or samples per hour, is not how long it takes for a sample peak to emerge after injection. Throughput is the number of samples that can be injected per hour. The delay time, or how long it takes a sample to pass from injection to the detector is a function of flow rate, and tubing volume.

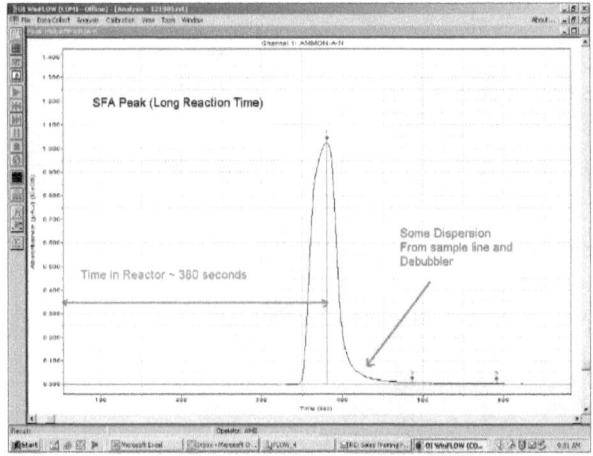

This is an ammonia peak by SFA after a dwell time (time in the cartridge) of 6 minutes. If this were FIA, the 6 minute dwell time would result in a very broad peak. The segmentation minimized dispersion. There is some tailing that is a result of de-bubbling just prior to the flow-cell. If this peak had been collected on an SFA system that did not de-bubble, the peak would be narrower and throughput possibly higher.

In a slow reacting method, FIA peaks exhibit high dispersion. The high dispersion causes FIA peaks to broaden decreasing the throughput. The air segmentation of SFA minimizes the dispersion keeping the peaks narrow.

The FIA sample travels through the tubing without air segmentation. FIA peaks are Gaussian shaped as a result of dispersion. FIA methods control dispersion to obtain adequate mixing and peak response. With fixed volume injection, an ideal FIA peak is Gaussian, and response can be measured by height or area, whereas SFA is not an absolute injection volume (controlled by time and size of the pump tube) and can only be measured by height. Assuming there is even flow, all FIA peak shapes will be similar resulting in almost no significant difference if measured by height or area. FIA methods do not attain steady state meaning that the analyte rarely reacts with the entire reagent. In SFA, the idea is to inject enough sample volume so that maximum reaction (steady state) is reached. As mentioned previously, most FIA methods are designed to get sample in and out of the cartridge in about a minute or less. A Gaussian peak shape is the goal. Samples in the cartridge for too little time result in leading peak shapes, and samples in for too much time become broader.

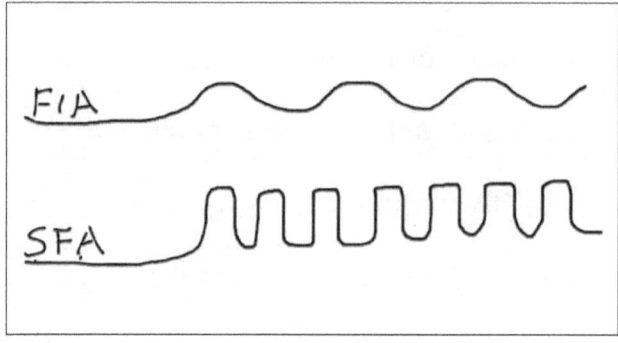

In this illustrated example, SFA peaks approximately double the FIA peaks in the same amount of time. Throughput is the number of samples that can be injected per hour. In this example, the high dispersion of the FIA method broadens the peaks decreasing the throughput. The air segmentation of SFA minimizes dispersion, keeps peaks narrow, and maintains high throughput.

What does all this dispersion stuff mean? Not much really because most environmental methods react in sufficient time that a FIA analysis can be completed in about 60 seconds or less. Unless you are doing an on-line digestion or very complex chemistry, either SFA or FIA is sufficient. If the reaction is complex, SFA should be used.

Reaction times

The reaction time is how long it takes a color reaction to produce a maximum Absorbance. The reaction time is a function of reagent ratios and temperature. Reagent ratios are experimentally determined during method development and should be kept constant. In continuous flow analysis reaction times are decreased by raising the temperature.

In a plot of absorbance versus time with a

vertical line drawn at one minute, if the maximum color is reached a FIA method that allows the sample to react for one minute will have reached maximum color, or steady state. In this chemistry the choice between FIA or SFA would not matter. Since FIA is simpler to operate, the logical choice would be a FIA method. An example is the generation of HCN by acidification.

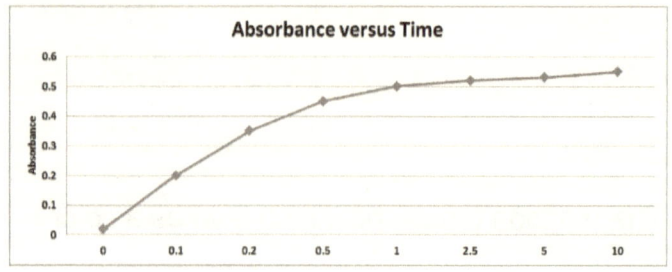

If however, the reaction is slower and at one minute has not reached maximum color, then analysis would not be at steady state. Assume the absorbance rises with time reaching a maximum at about five minutes. For this chemistry, you would want SFA so the reaction can proceed for five minutes (sample in cartridge for five minutes) and maximum response is attained. Of course, since SFA samples are injected, one right after another, the sample peaks could still be a minute apart even though it took five minutes for the first one to pass through. For a slow reaction SFA is preferred, or another option is to increase the heat for FIA analysis.

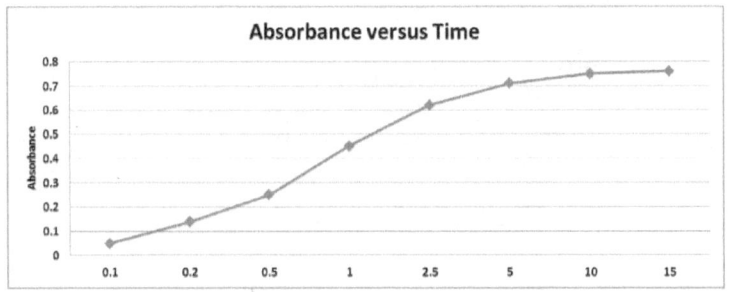

The same chemistry at increasing temperatures speeds the time to maximum Absorbance. For example, FIA can be used if the sample and reagents heated to 50 degrees increases the signal to a useable response. Method development should take into consideration potential undesirable reactions at elevated temperatures. As the color reaction rate increases with temperature, so do interferences.

Some similarities and differences between SFA and FIA

Flow injection analysis, FIA, measures a precise volume of sample at exactly the same time, every time. Peaks are Gaussian shape, not reaching steady state (or a flat top). The response is very reproducible with %RSD often below 1 even at very low concentrations. Noise is very low due to no air segmentation. The low noise results in high signal to noise ratios even if the overall maximum color development is not as high as SFA. Low detection limits are possible because of the low noise. Statements often made about SFA having lower detection limits than FIA due to SFA's ability to attain a steady state, or maximum Absorbance signal, are not true. The lower detection limits seen by some SFA methods are usually due to the peak broadening when FIA analyses of slow reacting chemistries are attempted.

Flow Injection analysis measures a sample reaction at the exact same time, every time, and assumes that all samples react at the same rate as the standards. This assumption is not necessarily valid. If the rate of reaction varies by matrix, then measuring before steady state is obtained could result in significant error.

If analysis of samples of widely varying matrices is needed, then it may be better to analyze by SFA and ensure that a steady state signal is reached. This will help minimize error that can result as reaction rates change due to matrix effects.

FIA analyzers inject reproducible volumes of sample by using fixed volume sample loops. FIA precisely controlled injections enable quantitation by peak area. SFA analyzers control sample injection volume by aspiration rate, and immersion time; an aspiration rate of 1 milliliter per minute immersed in the sample for 30 seconds will inject 500 micro-liters of sample. Between SFA injections the sample line pulls air into the cartridge as the sampler probe travels from the sample cup to the wash station and back. A reproducible SFA sample volume assumes all sample cups are filled precisely to the same level. Since they are not, there are always slight variations in SFA injection volumes. As long as the reaction is brought to steady state, these small variations in volume do not matter. However, because the absolute amount of sample injected is not constant, SFA quantification should be done only by height and not area.

Flow injection analyzers do not require that auto-sampler wash solutions match the sample matrix. Flow injection carrier solutions should match the sample matrix if it differs significantly from reagent water. For instance, the carrier for colorimetric cyanide after distillation should be 0.25 M NaOH to match the samples. The sampler wash solution can be reagent water. The sample wash solution is the carrier solution for segmented flow analysis and, thus, should closely match the sample matrix. If samples are preserved with sulfuric acid, then the wash should be sulfuric acid. If running colorimetric cyanide, the sampler wash should be 0.25 M NaOH.

Interpreting a Flow method diagram

The flow diagram tells the user how to assemble the cartridge and where to hook up reagents, the pump tube sizes are coded by color and the various sizes, or internal diameter, set the flow rate of each reagent. Starting from top to bottom on the cartridge:

- For SFA the first reagent should contain a surfactant. The surfactant "wets" the inner walls of the conduit tubing ensuring that segmentation gas has a slight cigar shape and that air/liquid slide easily through the tubing without surging.
- The sample, in SFA, sample volume determined by time in the sample solution and the size of the pump tube, in pumped into the first, usually a buffer or complex reagent solution. The sample solution and the wash solution alternate through this tube according to the timing of the auto-sampler. All other tubes are continuously pumping reagent. A mixing coil allows the sample and complex reagent to mix.

- The first color reagent is merged in and mixed.
- Additional color reagent may be merged in and mixed.
- The solution may travel through a heater to speed the reaction. Though SFA is capable of one-hour reaction times with minimal dispersion, this delay time is too long for reasonable lab work. Therefore, when heat is applied the internal volume of the heater should heat the sample at sufficient temperature and delay time so that sample reacts for no more than 5 – 10 minutes.
- The reacted, color developed, sample emerges from the heater into a de-bubbler that removes the segmentation gas just prior to the flow-cell. Some SFA systems de-bubble electronically, allowing bubbles to pass through the flow-cell. This minimizes

dispersion from a physical de-bubbler.
- The sample enters the flow-cell where it passes through wavelength specific light.
- The absorbance of light is proportional to concentration of ammonia in the sample.
- The signal output is a series of peaks corresponding to sample solution injections and baseline corresponding to wash solution containing no sample.

Summary of SFA and FIA

Segmented Flow Analysis allows methods to have longer reaction times and is useful when analyzing samples of widely varying matrices. The longer reaction time is simply delay between injection and the emergence of the first peak and does not affect throughput. In fact, for slow reacting analytes, SFA has a higher throughput than FIA. The auto-sampler wash solution is the carrier solution for SFA requiring that it match the sample matrix as closely as possible. Since SFA injects sample by aspiration time the slight variations in sample volume in the sample vials result in slight variations of sample injection volume. As long as SFA methods are allowed to attain a steady state signal the slight variations in injection volume do not matter, however, this does mean that SFA cannot be quantified by area and is always measured in height.

Attempts to increase or decrease sensitivity by changing the diameter of the sample aspiration tube, or reagent delivery tubes alter the chemistry of the entire SFA method. Increases in length of reaction coils, or mixing tubes, have little effect on the chemistry.

Flow injection analysis methods work best if the sample can be injected and detected in about one minute or less. This is possible with almost all environmental methods. Since FIA does not inject air, the apparatus is simpler and there is less noise. The minimal noise results in signal to noise ratios is almost equivalent to SFA, even if the maximum Absorbance of the SFA peaks are higher than FIA. There are no segments flowing through the FIA cartridge making it difficult to see the flow of reagents. This complicates troubleshooting slightly. To troubleshoot flow for FIA you should inject dye solution or introduce small air bubbles into the reagent lines. For SFA and FIA reagents should flow smoothly with a slight pulse. Do not over tighten platens on adjustable pumps; make them only as tight as necessary for a smooth flow.

Since FIA peaks are significantly affected by small changes in mixing and reaction tube lengths and volume, only experienced users should attempt method modifications. Method sensitivity is often easily modified by simply increasing or decreasing the volume of the sample injection loop.

The Discrete Analyzer

The discrete analyzer is an automated batch analyzer. Discrete analyzers more closely resemble manual methods than continuous flow analyzers. A discrete analyzer automates simple color reactions by pipetting and/or diluting sample, accurately adding reagent and mixing, incubating, final measurement, calculation and reporting of results.

A discrete analyzer is selective and only measures the test required for a particular sample and not every test. If you imagine a continuous flow analyzer with multiple channels running at once you can envision that it is running every injection for every test whether you need that test on that sample or not. A selective discrete analyzer allows the analyst to place samples in any order on the sample tray and then by software choose only the tests needed for that sample.

A discrete analyzer measures the analyte in any order in the sequence (or schedule) on any sample without possibility of carryover. Again, imagine a continuous flow analyzer. It is configured to analyze tests based on the chemistry mixing manifolds and the reagents that are pumped into it. It is not possible for a continuous flow analyzer to run one or two samples for one test, automatically reconfigure itself with a new wavelength and different reagents and immediately follow with another test.

Also, even if an analyst changed the pump tubes and the filters, the cartridge would need to be washed out to remove the remains of previous testing. A random-access discrete analyzer reacts and reads samples in self-contained cuvettes.

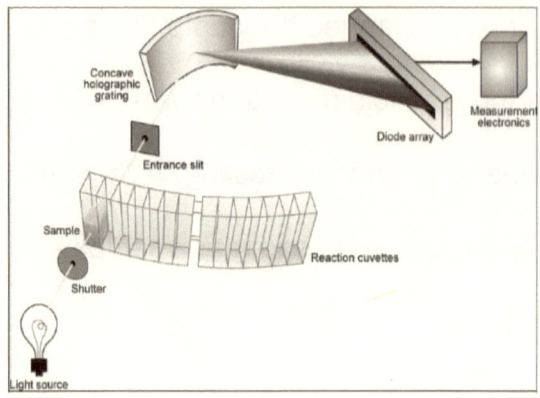

A discrete analyzer is a true random access analyzer. For an auto-sampler, random access means the probe can go to any position on the x, y axis. Continuous flow analyzers are not selective meaning that samples must be pre-selected according to the tests required. A flow analyst gathers samples by test, and runs the instrument according to the test that was preconfigured.

A selective analyzer only measures the test required for a particular sample and not every test. If you imagine a continuous flow analyzer with multiple channels running at once you can envision that it is running every injection for every test whether you need that test on that sample or not. A selective discrete analyzer allows the analyst to place samples in any order on the sample tray and then by software choose only the tests needed for that sample.

True random access means that the analyte is determined in any order in the sequence (or schedule) on any sample without possibility of carryover.

A discrete analyzer only analyzes the method needed per sample. This method changeover occurs automatically, and the methods needed per sample are selected by the operator when creating the analytical schedule prior to sample analysis. The analyst walks away, and the instrument commences to run the samples for only the tests requested.

When discrete analyzers were first introduced, computer technology was not capable of handling the programming requirements for ready operation of the instruments designed. Now that computing power is sufficient, and computers are inexpensive, discrete analyzers are becoming more viable. Discrete analyzers maintain the integrity of the sample by keeping the entire reaction constrained in its own receptacle contrary to flow (or even chromatography) methods where once injected the sample is "lost" in the stream and only identified by time. Discrete analyzers automate manual methods by delivering samples, standards reagents and diluents with a robotic arm (instead of pump tubes – flow methods). Reactions occur in individual cuvettes with only micro-liters of sample and reagents. After incubation the absorbance is read in the reaction cuvette. The entire process is controlled by computer with high precision and accuracy. Current discrete analyzer methods are being written as adaptations of manual and/or flow methods.

The basic discrete analyzer operation dispenses, mixes, reacts, and measures an analyte all within the confines of a reaction cuvette.

Comparison of Discrete and Continuous Flow Analyzers

In a continuous flow analyzer reagent is always flowing down the path of the tubing. Samples are injected into the reagents, but there are also sections of tubing that contain reagents without sample. Since discrete analyzers measure samples contained in individual cuvettes there is no stream of flowing reagent, therefore, a discrete analyzer only consumes the reagent required for that test and no extra. Carryover, or the interaction of one sample with the one following it as it travels down the tube, is not applicable to discrete analyzers. Carryover is minimized in flow analyzers but is always there to some degree. Throughput in discrete analyzers is a function of the number of reagents per test. The more reagents to dispense the longer the test will take. In flow analyzers the reagents are continuously injected into a flowing stream. The throughput of a flow analyzer is defined by peak shape and width; the wider the peak the less the throughput.

The operator maintenance of a discrete analyzer is limited to a very minimal changing of wash water, and very rarely changing of things like lamps, wash station tubing, probes and so forth. Continuous flow analyzers require daily to biweekly changing of pump tubes, and the interchange of cartridges when switching an analyzer from one method to the next. The time to switch methods depends on the analyzer and the experience of the analyst. An experienced analyst can easily set up or tear down a CFA method in about 5 minutes. Major repairs on a discrete analyzer require an experienced technician, or even retooling at the factory.

A flow analyzer can operate in continuous mode (samples can be continuously added while the instrument is running) while the discrete analyzer is designed for batch analysis only. Some discrete analyzers allow continuous feed of reaction segments and loading of samples; however, these functions tend to add mechanical complexity. Continuous feed on a flow analyzer does not add complexity because the samples and reagents are always flowing in a continuous stream through the detector.

A flow analyzer is not selective. All tests configured to run upon sample injection are done whether you want them or not. A flow analyzer is not random. You cannot select tests by sample container, instead all tests are run each time the flow analyzer samples from that vial.

A discrete analyzer selectively runs only the methods chosen by the software and only on the sample vials that those specific tests were requested for. This is a fundamental difference between discrete and flow analyzers. While most flow analyzers have random access auto samplers, meaning that the software chooses which sample vial to sample from, this is not the same as only analyzing the tests chosen for that particular vial.

Flow analyzers are "fast" analyzers. Sample throughput is defined by the width of the sample peak and not by the number of reagents. A 6-channel flow analyzer injecting one sample per minute is running at 360 tests per hour.

Discrete analyzers are single channel instruments with throughput being limited by the number of reagents added per test. This makes them fundamentally slower than continuous flow. In the future, as we begin to develop discrete analyzer methods instead of modifying flow analyzer methods to discrete analyzers, I suspect throughput will be eventually increased. For instance, ammonia is four-reagent chemistry by flow and can be made into a very effective two reagent chemistry on discrete analyzers.

Choosing between Continuous Flow or Discrete Analyzers

The technique that you decide to use to automate your lab is your choice. However, this choice should be made on what is best for the laboratory after a careful examination of the options. Sometimes it may be best to modify your thinking and redefine laboratory operations to make full use of the technologies that exist. It makes no sense to operate a discrete analyzer like a flow analyzer because it is not one. It also makes no sense to arbitrarily decide on one technique above another based on "personal preference". Not if the goal is to analyze more samples and increase profit. Automation techniques need to be chosen based on what works best, not just in your lab, but for the analyte as well.

Many functions that the analyst controls on continuous flow analyzers are computer controlled on the discrete analyzer. These include the number of tests run per sample, the number of reagents that can be added per test, the transfer and mixing of sample and reagent, blanking of individual samples, and the analytical wavelength. Dependence on computer control makes the discrete analyzer more complex, and also limits expandability of the hardware to the original instrument design. It is easy to add extra functions, such as distillation or extraction as modules to the CFA instrument after an initial purchase. These functions, if they even exist at this time, would require a factory installation onto a discrete analyzer.

Different Types of Discrete Analyzers

Discrete analyzers can be divided in various ways. Of the current commercially available two divisions stand out. Those discrete analyzers that use a flow-cell and those that measure in the reaction cuvette; and those that wash the cuvette versus using disposable cuvettes.

Flow-cell designs

The flow-cell designs have become popular in environmental testing discrete analyzers from the ability to add a longer path cell in efforts to decrease the detection limits. These analyzers dispense and react into individual cuvettes and then transfer the reacted sample solution to a flow through cell for final absorbance measurement. Manufacturers claim low detection limits with these analyzers. Flow-thru cell designs require rinsing with several cell volumes between samples to ensure that all of the previous sample was rinsed out prior to introduction of the next sample. While sample integrity is maintained during the color reaction step, it is lost upon transfer of the sample to the flow-cell. The risk of sample to sample contamination is not the risk of the contents of sample a contaminating sample b, but the risk of reagents from sample a carrying over into sample b. This risk of reagent contamination limits the ability of flow through cell discrete analyzer designs to rapidly switch from one method to the next. The only way minimal cross contamination of reagents from one test to analyte of the next test can be ensured is long rinse cycles in between tests. This "problem" has led to the use of flow cell design discrete analyzers as single test sequential analyzers; basically as a single channel continuous flow analyzer without pump tubes.

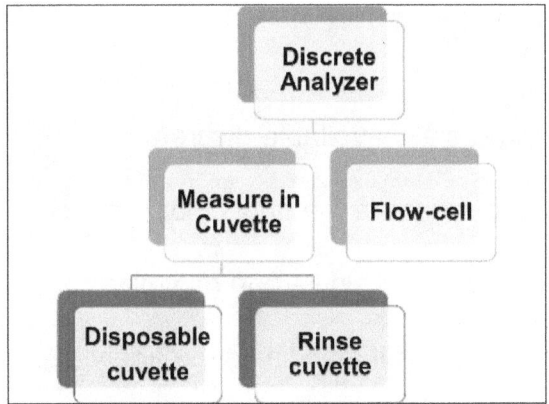

Of the current models available, sample absorbance is measured in the reaction cuvette, or the reacted sample is transferred and absorbance is measured in a flow-cell. Most of the read in the cuvette designs use disposable cuvettes; however, some analyzers automatically wash and re-use them.

In reaction cuvette designs

Other commercially available discrete analyzers react and measure the final color product within the reaction vessel. This design minimizes carryover contamination but also limits the analytical path-length to the path defined by the manufacturer during instrument development. Some of the read in cuvette designs wash the cuvette between tests. While the cuvette ensures sample integrity is not lost during reaction and measurement, the washout step cannot guarantee that all reagent is rinsed out between tests. Thus, similar to flow-cell designs, washing cuvettes risk contamination of major ingredients of the reagent if that ingredient is an analyte in the next test. This "washout" requires long rinse cycles between tests limiting throughput. Maximum throughput between tests can only be obtained by using disposable cuvettes and a read in the reaction cuvette design.

For more information on CFA or Discrete Analyzers Google:

Ewing's Analytical Instrumentation Handbook, Second Edition

Ewing's Analytical Instrumentation Handbook, Third Edition

Automatic Chemical Analysis by Peter B Stockwell

Ion Chromatography

In some ways, Ion Chromatography is similar to flow injection analysis. The sample is injected into a flowing stream of reagent within small diameter tubing using a valve. The purpose of ion chromatography is to separate analytes and measure them separately, however, while the purpose of flow injection analysis is to mix the sample and reagents together and measure a reaction product. With ion chromatography it is possible to separate ions, react them chemically, and determine the concentration of individual reaction products from a single sample injection. Thus, with ion chromatography, multiple analytes can be determined from one injection.

One of the most common uses of ion chromatography is the determination of anions in aqueous samples. With one injection the most common anions in water can be accurately measured in about 10 to 15 minutes. Anion chromatography has been used extensively in the analysis of anions and disinfection by products in Drinking Water.

The typical ion chromatograph consists of an auto sampler, a pump, an injection valve, a column and guard column, a suppressor, and a detector. Separation of analytes takes place on the column while the guard column helps to prevent contamination of the column. The suppressor converts the analyte to a form more readily detected by the detector.

Ion Chromatography is an excellent technique for the determination of anions in drinking water, and clean groundwater and wastewater. In samples of unknown matrices conductivity should first be determined so that dilutions can be made prior to analysis. These dilutions are necessary to avoid overlap of large peaks with smaller peaks and to bring large amounts of analyte (usually chloride or sulfate) within calibration range. Unfortunately, this necessary pre-dilution sometimes causes other analytes to be diluted below detectable levels.

A chemical method such as SFA, FIA, or Discrete rarely requires pre-dilution to remove interferences. In Ion Chromatography large chloride peaks will overlap nitrite peaks making determination difficult but in chemical methods chloride does not interfere.

In ion chromatography large sulfate peaks can cause retention time shifts and actually be measured as phosphate. The sample must be diluted and reanalyzed for sulfate, but this causes phosphate number to be lost. In chemical methods sulfate does not interfere with phosphate.

Acidic samples and samples containing high concentrations of trace metals should not be analyzed without dilution. Acidic samples, besides having large amounts of the acid anion, cause retention time shifts resulting in misidentification of analyte peaks. Trace metals will occupy active site on the column eventually ruining its ability to separate the ions. Chemical methods are typically developed to avoid interferences from acidity and trace metals. For instance, low levels of chloride can be determined chemically in the presence of very high concentrations of iron, aluminum, copper, and sulfuric acid.

Although ion chromatography performs well analyzing nitrite in drinking water, the short holding time, in un-preserved samples, requires that chromatographic runs be set up specifically for the analysis of this ion before its holding time is expired. If chloride or sulfate are off scale the sample must be rerun again. Chemical methods are more suited to quickly analyze short holding time parameters such as nitrite and phosphate.

Interferences with ion chromatography

IC or ion chromatography separates ions from each other using exchange resins. During the exchange process positively charged ions, or cations and metals, are pulled from solution and replaced by Hydrogen ions. The resulting hydrogen anion, or acid, is detected by a very sensitive electrochemical detector.

The identity of the ion is determined by retention time, and the concentration is proportional to the intensity of the detector output.

Accurately determining ionic species by IC depends on the efficiency of the separation of the ions from themselves and from the matrix. Wastewater matrices often contain anions in concentrations very different from each other. For instance, it is not uncommon for chloride to be several hundred to thousands of ppm higher than other ions of interest (nitrite). This means that several determinations must be made per sample to obtain results for every analyte within their calibration ranges. Also, since ions must be resolved from each other, it is often necessary to dilute a sample to get rid of interferences but doing so raises the detection limit of analytes that are already near their detection limit to a value of no use to the end user.

Advantages of ion chromatography

IC determines anions and/or cations accurately in a single run. Assuming all results are on scale every anion can be quantified in less than 15 minutes. IC is approved for EPA compliance reporting of cations and anions in drinking water and wastewater.

IC is extremely useful when determining every

anion present in the sample solution and is the best way to determine sulfate in 'normal' waters. Many laboratories use IC for chloride and sulfate in hundreds of wastewater samples per week but are still unable to use IC economically for all samples, or for other anions, mainly fluoride, nitrate, nitrite, and phosphate.

Ammonium:

- Ammonium is a cation and cannot be analyzed using the same method/run as anions. It is possible to configure an IC to run ammonium ion (and the other cations plus a few metals) and all the anions at the same time from the same injection. This, however, means that a single ammonia result takes about 15 minutes per sample.

Fluoride;

- The water in the sample itself can interfere with fluoride on the IC. Newer instruments and columns claim to have solved this.
- Fluoride elutes first in an IC run, if fluoride is the only analyte of interest it makes more sense to use the CFA ISE method than IC than to make a 12-minute run for a single analyte.

Chloride;

- Chloride elutes immediately after fluoride.
- IC determines chloride with great sensitivity and accuracy. This sensitivity allows lower detection limits, but also increases chloride's ability to interfere with nitrite, the next anion to elute.
- Most samples require dilution for an

accurate chloride determination. Many times, multiple dilutions must be made just to get results on scale. These multiple dilutions can increase analysis times to an hour to obtain a single chloride result.

- If a laboratory is analyzing multiple samples for chloride only it is more cost effective to use a wet chemistry or CFA/ISE method.

- Samples high in trace metals, acidity, or cations will eventually deteriorate the IC column requiring early replacement. IC is difficult, at best, when determining chloride in acid runoff.

Nitrite

- Elutes immediately after chloride.
- High chloride concentrations can overlap

the nitrite peak, requiring a dilution. Dilution sometimes raises the MDL beyond usefulness. Using a tandem conductivity and UV detector enables determination of nitrite in high chloride solutions.

- Nitrite has a holding time of 48 hours in unpreserved samples requiring samples to be analyzed as soon as possible once received.
- An IC run is 12 – 15 minutes. If samples are being analyzed for nitrite alone this results in ~ 12 minutes per test. CFA methods can analyze nitrite at up to 100 samples per hour.
- The startup time for an IC is 30 minutes to 1 hour. A series of nitrite samples could be analyzed on a discrete analyzer before the IC has even stabilized.

Bromide

- Unless only running bromide, IC is the best method.

Nitrate

- Nitrate detection limits by IC are similar to flow on clean samples.
- Saline samples that require dilution force IC detection limits higher.
- Nitrate and Nitrite analyses by IC must be done within 48 hours if samples are not preserved.

Phosphate

- Phosphate elution is pH dependent; improper pH of samples or eluent can cause phosphate retention times to shift.

- Detection limits are often lower by colorimetric methods.
- Phosphate should be analyzed within 48 hours.
- A series of phosphate samples could be analyzed on a discrete analyzer before the IC has even stabilized.
- IC cannot easily analyze total Phosphorus. Also, IC runs for total P are 5 – 10 minutes per sample compared to about 1 minute by ACA methods.

Sulfate

- Sulfate is the last major anion to elute. Sulfate works very well by IC other than analysis time.
- A Method 300 analysis time is about 15

minutes. If a laboratory is analyzing multiple samples for sulfate only (or chloride and sulfate only) ACA methods are more rapid, increasing throughput.

- Many samples require multiple dilutions for an accurate sulfate result. These dilutions increase analysis time to nearly an hour to obtain a single sulfate result.

Summary of ion chromatography

To use IC economically requires that more than one anion be desired. Using IC for a single analyte, such as ammonia, is not economical since faster methods that can handle more complex and preserved matrices are readily available. Wastewater analytes with short holding times, such as nitrate, nitrite, and phosphate inconvenience the laboratory running these analytes by IC since they are inevitably almost out of holding when they arrive at the laboratory. In contrast, the ease of use of discrete analyzers makes the rapid analysis of nitrite and phosphate easy. Many labs use IC to analyze chloride and sulfate. If there are a lot of samples a simultaneous flow instrument is capable of increasing throughput from 4 samples per hour by IC to 60 samples per hour by flow (15 times more samples per day).

Conclusion

This has been a short introduction to wet chemical analysis. This book is a compilation of notes from presentations I have done. I hope you have found it useful. Please visit my web page at www.williamlipps.org for more information on chemical analysis, and environmental methods.

Thank You for reading.

www.ingramcontent.com/pod-product-compliance
Lightning Source LLC
Chambersburg PA
CBHW020907180526
45163CB00007B/2651